About the Author

Lisa Falco has been working in the interface between medicine and technology for almost 20 years, on how data can be used to better understand the
human body. Doing so she has worked within varying fields of women's health,
such as bone research, diabetes and fertility. She was previously Director of
Data Science at Ava, a Swiss FemTech startup, where she led a team developing
algorithms and artificial intelligence in order to help women get pregnant faster
by using their physiological data based on their menstrual cycle. The work she
did there, and the women she encountered in the process, inspired her to write
this book.

Published by Clink Street Publishing 2021

Copyright © Lisa Falco, 2021

Content editing: Barbara J Müller / Design and graphics: Patricia Wassmer

ISBN 978-1-913962-24-1

GO FIGURE!

The astonishing science of the female body

Lisa Falco

All the facts in this book are collected to the best of the author's ability and the scientific or literary source of the knowledge should always have been cited. If a source has fallen out in the editing process, this is an honest mistake that should be corrected for future editions. If you spot such a mistake, please don't hesitate to inform the author.

To my family, close and far

TABLE OF CONTENTS

INTRODUCTION

The female body and mind are often portrayed as a mysterious temple that is impossible to understand. With this book I want to unravel some of the mysteries surrounding women's biology and explain what is happening underneath the surface. With organisms as complicated as the human body, it is necessary to have a broad knowledge to understand how everything is interconnected. This book combines that overview with the captivating nitty-gritty details that demonstrate the ingenious construction of the female body, debunking a few myths along the way.

We all know that the female body changes cyclically every month during the reproductive years, and that it completely transforms during the major transition phases: puberty, pregnancy and menopause. However, most of us are not aware of the fascinating details. What triggers those changes, and what are the sometimes-unexpected consequences?

You will also learn many unexpected things:
- How pregnancy is a temporary hijacking of the body and how this changes not only the body, but also the brain, everything directed by a new organ growing inside of it.

- Why women get more anxious as mothers and when they grow older.
- That men can breastfeed if given the right hormones.
- That vision and voice changes with the menstrual cycle.
- That women carry parts of their children (actual cells) with them forever after giving birth.
- How fertility can be even more complex than you thought.
- That sperm is being very well taken care of and are not at all the brave little swimmers that we always thought.
- Whether the mood swings before menstruation are real or a social construction (spoiler: they are real!).
- Why women are more robust towards infections than men while still suffering more frequently from autoimmune diseases?
- How does a regular menstrual cycle protect heart and bones?
- What happens when you age?
- Do menstrual cycles really synchronize?

This book answers these questions and many more. It is based on science, but you should not shy away as I guide you gently through this fascinating journey. You can choose your itinerary since each chapter can be read independently. Our aim: comprehension, empowerment and fun. Whenever your memory needs a refresher, there is a list of the most common medical terms at the end of the book.

A body in continuous transformation

One thing that makes the woman's body so extraordinary is its continuous changes. You not only have the major transitions – puberty, pregnancy, post-pregnancy, perimenopause and menopause, – but also the cyclic transformation happening each month.

The hormonal changes over the menstrual cycle can impact our mood, how we react to pain, how hungry we are, how strong our immune system is, and what our bodies look like, just to name a few examples.

For some women, the changes over the menstrual cycle can be a radical experience with many ups and downs, whereas other women can go through the cycle while hardly noticing any impact. The major transition phases are hard to go through without noticing. In puberty, oestrogen changes the body's shape and function fundamentally. The breasts develop and the levels of fat increase under the skin, on the hips, thighs, and buttocks, all that accompanied by emotional storms, changing body odour, and hair growths. Everything is triggered by the new hormones.

Pregnancy might be the largest transformation of them all. The body is turned into a machine capable of producing and nourishing life. It is easy to observe the outer changes, such as the growing belly, the increasing size of the breasts and the additional fat needed to provide food for your growing baby, both inside and outside of your womb. Again, these changes are triggered by hormones, but this time released by the placenta, a new organ (you read correctly) formed inside the womb in the early pregnancy. On the inside, the placenta is not only releasing hormones, but also acting as a life support machine, providing the developing foetus with oxygen and nutrients while cleaning the foetus' blood. At the same time the breasts are growing and changing their entire internal structure to be able to produce milk for the new baby.

The last major transformation in a woman's life is the menopause. This is when her menstrual cycles stop once and for all. It marks the end of her reproductive life, and this is a transition that is dreaded by some and celebrated by others. The sudden removal of sexual hormones has a large impact, not only on the reproductive capacity, but it also affects metabolism, bone and muscle strength, ageing of the skin, and mental health. It might sound terrifying but it is a very natural part of growing older and a key to the successful evolution of humankind.

Later in this book you can read about the captivating details driving these transformations. This will allow you not only to observe what is happening, but also to understand it. Hopefully, this will help you feel more in control and able to enjoy those changes.

Understanding ourselves and others

We are often asked to believe and accept things as they are, without really understanding them. However, understanding is a cornerstone of tolerance, it is difficult to be tolerant about things we cannot relate to, or find logic. Women are often seen as difficult to understand, sometimes even by ourselves. This lack of understanding can in many cases lead to a lack of respect for our experiences, whether it is related to our physical or mental wellbeing.

A close friend said to me the other day: "This thing about PMS (premenstrual syndrome), isn't it just an excuse for women to be bitchy?" I could not believe how she as a woman could say such a thing, but of course, she had never experienced PMS herself and could not relate to how it felt. This gave her the feeling that the whole thing was exaggerated.

Personally, I have no doubts that PMS exists. I never become 'bitchy', but I come home almost crying from work once a month, firmly convinced that everyone hates me, and that I am the most useless person on the planet. One day later, these feelings are gone and two days later, I get my next period. Same procedure, every month. Either PMS is true, or the world has synchronized itself with my menstrual cycle.

The purpose of this book is to try to explain in detail the physiological processes inside our bodies that influence us. One such subject is PMS, and the chain of events that may lead to depressed and anxious feelings. Hopefully, not only my friend, but also everyone else reading this book will get a better understanding of the experience many women have and thereby become more tolerant towards it.

I must admit that earlier in my life, I also had a problem respecting other women's experiences. One in particular was the experience of menstrual cramps. I did not experience them myself, so somehow, I thought women who were incapacitated with pain were being overly sensitive. Unfortunately, in the past my own ignorance was shared with many medical professionals, and women's pain has been ignored for much too long. Nowadays, we understand better which underlying issues cause the excessive pain. This both increases the tolerance for it, and helps women get treatment faster, whether it is PMS, period pain, emotional storms in puberty or hot flashes in menopause.

If you are a woman planning to get pregnant or experience problems with your health, it feels obvious to take an interest in your body and learn as much as possible. However learning about your body is relevant in all life stages. Your body will accompany you throughout your whole life, and many of the choices you make when you are young will have an impact later.

We are not victims of our hormones and our bodies, and we also should not exaggerate their effect. That being said, knowledge is power and being mindful about what is happening inside us can help us to either embrace it, or to put it aside and concentrate on other things.

Just Another Word Before We Start

The difference between you and science

What you will read in this book is based on science, but what does that really mean? A lot of medical science is based on clinical trials, where a group of people or animals are studied under different conditions. Clinical research can often seem a bit dodgy, and there are a few things that are important to know when you read about it. Mostly, the results are never black or white or straightforward to interpret, which is why statistics is needed.

Statistics can be seen as a 'dark' art before you get the hang of it. One joke about statisticians goes: "A statistician can have his head in the oven and his feet in the freezer and say that on average he is fine." Statisticians often talk about averages, but it is also important to look at distributions. If we look at the statistician with the feet in the freezer, he will cover the entire distribution ranging from very cold (the feet) to very hot (the head). If you now imagine the statistician with a very large belly and with a small head and small feet you would get a good impression of a distribution. Only a very small portion of the body is at the extremes and the largest portion of the body is somewhere in the middle and experiences reasonable temperatures. This is a typical normal distribution; most things fall somewhere in the middle and a few things in the extreme areas. Very few people are exactly average, but most people fall within a distribution and it is the limits of the distributions that defines what is considered 'normal'. The head and feet of the statistician would in this case not be considered 'normal'. For instance, on average, women have a menstrual cycle that lasts 29 days, but it is completely normal to have anything between 24 and 35 days.[1]

Another important concept in statistics is that *correlation* is not *causation*. This means that just because researchers have managed to see a link between two phenomena, it does not mean that one causes the other. A famous example is the correlation between chocolate consumption and Nobel Prizes. In a study from 2012 it was shown that countries with a high consumption of chocolate produced more Nobel Laureates.[2] The correlation between the two is very strong but that does not mean that there is a causation. Despite dark chocolate having many beneficial effects on your health, the most likely reason between chocolate consumption and Nobel Prizes is that economically wealthy countries spend more money both on education and luxury goods such as chocolate. Switzerland, which is the country with the most Nobel Prize winners per capita also happens to be very famous for its chocolate and therefore bias the results.

It is also important to realize that many scientific results do not have a direct impact on you as an individual. Many of the findings from clinical studies apply to a large population. It means that the results matter on a large scale but not necessarily to you. One example is the impact of breastfeeding. It has been shown that on average, breastfeeding increases the IQ score of babies by 3 points.[3] First of all, it is very unlikely that you will be representative of the average effect. If the average increase is 3 points, it means that for your child, the effect of breast-feeding can be anywhere but probably somewhere between 0 and 6 points. Even though Einstein was allegedly breastfed until he was six years old, this was most likely not the reason he was so smart.

However, let us assume that you would be representative of the average effect. Your child will improve its IQ score with 3 points if you breastfeed it. An increase of 3 IQ points for your child will not bring it any additional advantages in life; for that, the difference is too small. However, on a society level it makes a big impact if the population has an IQ score that is 3 points higher on average, which is one of the many reasons it makes sense to promote breastfeeding.

I am a strong believer in science, but I remain aware of its many flaws. Despite that, I have just like most other people, a tendency in believing what confirms what I would like to believe in. This is called *expectation bias.* If you expect some-thing to be true, or want it to be true, it is much more likely that you will find indications that confirm your beliefs. You also more likely refer to research results supporting your beliefs.

One such example is all the research supporting that a glass of wine a day is good for your health. That is a marvellous finding for a wine-lover like myself, so I am of course very prone to believe it and celebrate the finding with a nice glass of Cabernet Sauvignon. As it is not my domain of research, my own belief in the benefits of wine drinking is not harmful for anyone, except for myself in case it would turn out to be wrong. However, if I were researching the effects of alcohol, such a bias could be harmful.

An example of how bias has influenced research over the last centuries is the misogyny that has been directed towards women. This negative idea has pushed researchers towards findings demonstrating women's inferiority to men. Even prominent thinkers like Darwin used his theories of evolution to justify the idea that women were intellectually inferior to men. This was a belief so deeply rooted in society that he did not manage to see beyond it.[4] Luckily, things are evolving but it takes time to change such a well-established preconception, no matter how unfounded it is. As research in the past was mainly conducted by men, more focus has been on research on them, and women have not been studied properly. Having more women in science may lead to a better balance in the future.

It is easy to identify bias in the past, but it is difficult to say what type of biases we are carrying with us now. Trying to be aware of our own prejudices is just as important for researchers as it is to us as consumers of media and scientific communications.[5]

Another important thing to consider regarding clinical research is something called confounding factors. It is all the things that you have not examined, which might have an impact on your results and prevent you from seeing the real link between the input and output of the observations. This is also linked to correlation and causation, and it is basically what makes the difference between the two. If you have observed a correlation in your data, the confounding factors are what prevents you from finding the cause.

Again, wine studies are a good example of confounding factors and correlations. In many studies where the benefits of wine were shown, the participants were grouped depending on their drinking behaviour. People who drank nothing were in one group, people with moderate wine drinking (about one glass a day for women, two for men) in a second group, people who drank more in a third one. The studies could show that the group of people with a moderate consumption lived longer than, not only the group with a heavier drinking, but also longer than the group that did not drink any alcohol.

Can you draw the conclusion from this observation that wine drinking is good for you? Not really, many other factors influence the results. For example, there is a possibility that moderate wine drinkers have a higher socioeconomic standard, and such people tend to live longer. In this case, the researchers had done their homework and compensated for this, as well as many other factors. However, there was one thing that some of them had missed: in the group of people who did not drink alcohol, former alcoholics and people with chronic illness could have been included. As alcoholics and people with chronic illnesses have a shorter life expectancy, they would shorten the average lifespan of their group. Therefore, the conclusion that it is better to drink one glass a day than none may be biased.[6]

Temperature-box

Another fact that is even deeper rooted in our society is that our average body temperature is 37°C. Nowadays, that number has been adjusted to 36.6°C. The most probable cause is that we are much healthier and have less infections today, so our body temperature really is colder.[7]

Many truths have an expiry date. It is a common saying that a normal truth lasts 20 years and I suspect that this is correct, or maybe even shorter. Hopefully, most of the things you will read in this book will still hold in 10, 20 years' time, but you must always be prepared to challenge what you have learned in the past, whether from school or in this book.

Certain things cannot be studied for ethical reasons. One such topic is, staying on the wine drinking theme, the effects of moderate alcohol consumption during pregnancy. We know that heavy alcohol consumption exposes the baby to a very high risk of birth defects for instance maternal alcohol exposure during early and late pregnancy can cause both structural and functional defects in long-term renal function.[8] However, we do not know what would happen if you would only drink a little. It is also very likely that we will never know this. A woman with moderate alcohol consumption has no problem in stopping drinking for nine months and would therefore not take the risk to do any harm to her unborn baby. Nevertheless, such a study would never pass the various national ethical committees since the benefits of that knowledge is not in proportion to the risk the participants would expose their offspring to.

Most of the research cited in this book has been conducted in Europe and in the US. This might unfortunately bias some of the findings. Some studies have shown that there is a difference in what is considered 'normal' hormonal levels between women living in different areas of the world. For instance, studies of women in Bolivia have shown that they can conceive and go through healthy pregnancies at much lower hormonal levels than women living in Chicago.[9] That being said, the fundamental biology does not differ between ethnicities. In the US, where they have better possibilities than in Europe to conduct studies on diverse populations, most of the differences seen between ethnicities can sadly be traced back to differences in socioeconomic status and precarious living conditions rather than biology. One thing to bear in mind is that being an ethnic minority can induce a lot of stress and that stress can have an impact on your health and your menstrual cycle. I will therefore rarely mention differences in ethnicities but rather refer to the factor that might change the data, such as body mass index, diet and stress. Wherever I have found differences in the data, this will be mentioned. Lifestyle is another bias in the data where European and North American is considered the norm. Not because I think that they should be, but because the research I found was conducted in those countries. If I rewrite this book in ten years, I hope the input data will be more diverse.

I am primarily a scientific writer but to make very dry and complex data understandable and interesting, I will sometimes simplify descriptions while trying to remain as scientific as possible.

And with those words, let's get started on our learning journey. We start off by investigating what is a woman and how is she seen by science ...

WHAT MAKES A WOMAN?

For centuries, women were seen by science as men, but with pesky hormones. When my son was four years old, he had a similar opinion but the other way around. His explanation of a man was: *"It's like a lady, but with balls. You could say that it is a ball-lady [Hodenfrau in German]."*

For a long time, it was a joke in our family to refer to men as 'ball-ladies.' But just like my son has grown up to realize that men are not just women with testicles, science has now grown up to realize that women are not just men with odd or pesky hormones.

In which aspects are women so different? Many of the differences can indeed be traced back to the hormones which have a tremendous influence. People undergoing feminizing hormone therapy can immediately spot the changes that the hormones are doing to them. After three to six months of hormone therapy, the skin becomes softer, the muscle mass decreases, breast starts to grow, and the fat gets redistributed and accumulates around the hips and thighs.[10] As those hormones trigger such huge changes, it is no surprise that women can both feel and look different as these very hormones change cyclically and even more during the major transformations that they go through in puberty, menopause and during pregnancy.

What else makes women so extraordinary and differentiates them from the other half of the planet's population? At the time of writing, the definition of a woman on Wikipedia was as follows:

"Typically, a woman has two X chromosomes and is capable of pregnancy and giving birth from puberty until menopause. Female anatomy, as distinguished from male anatomy, includes the fallopian tubes, ovaries, uterus, vulva, breasts, Skene's glands, and Bartholin's glands. The female pelvis is wider than the male, the hips are generally broader, and women have significantly less facial and other body hair. On average, women are shorter and less muscular than men."

According to Wikipedia, the things that distinguish the two sexes are mainly the attributes linked to making babies, having babies and feeding babies. Even though being able to create life is definitely one of the big marvels of the female body, there is of course more to the story and one can be a woman without being able to get pregnant and give birth. What may seem like an easy and uncontroversial distinction at first glance is in fact a rather complex question. First, we need to distinguish between gender and sex. Gender is related to identity and sex to biology, but even if we only consider the definition of sex, it is not straightforward. You can be a woman without having all the attributes that are typically linked to female biology, and you can be a man while having some of those attributes. There is no clearly defined line, and what one would think was a purely biological question cannot be answered without being political.[11] The definition of sex is far beyond the scope of this book, in which the main issue is the effect of sex hormones. I will use the term 'woman' and 'female' in a broad sense without any attempt to classify individuals according to sex and gender.

Now, back to the pesky hormones, or the marvellous chemicals depending on how you see them, they are indeed to blame for a lot of things linked to the female biology. All the physical attributes in the Wikipedia definition have been created under the influence of hormones and they start playing a role even before birth. The very first job of the hormones is to turn women into men. Yes, you read that right. It turns out that my son's early view on the world agrees with our current knowledge, namely that the default human is female and becoming a man is an active deviation from the default.

Switch-box

When we are just a foetus, a few weeks after our mother's egg gets fertilized, there is no way to tell if we will become a man or a woman. Where there will later be our reproductive systems, there is just something called the gonads (or sex glands) and two systems of ducts, the Müllerian ducts and Wolffian ducts. Becoming a female means that the gonads will develop into ovaries, and the Müllerian tract will develop into the fallopian tubes, the uterus and the vagina. This will happen if there is no interference in the process. However, there is a kind of master switch, which is a special gene situated on the Y-chromosome. If this is present, it will turn the foetus into a male by developing the gonads into testicles. This happens about seven weeks into pregnancy. Once the testicles have been developed, they start producing anti-Müllerian hormone (AMH) and testosterone. As you might guess from the name, AMH's main job is to demount the Müllerian tracts; from the Wolffian tracts the connections between the testicles and other male genital parts are created. The clitoris and penis are formed from the same tissue but under the influence of sex hormones they develop in different directions.

Most of the external differences seen in men and women are secondary effects of these internal structures. The ovaries and testicles do not only produce eggs and sperm, but they also produce sexual hormones. The ovary produces oestrogen and progesterone, whereas the testicles produce testosterone. As already mentioned, these hormones have a huge impact on how the bodies develop and are the reasons behind many of the differences that can be observed in men and women when we reach puberty. Even though some hormones are strongly associated with men and others with women, both sexes produce all hormones but in different amount.

Asymmetry-box

For both males and females the gonad on the right side seems to be both heavier, larger and contains more proteins and DNA than that on the left side. This might be difficult to observe in women but easier to see in men. Despite this difference in size, there is the same number of follicles (what will later develop to eggs) in both ovaries.

Women are not the norm

Whether biology agrees or not, women are not the norm in society. Most things, from how the workplace is organized, to how public toilets are planned, are designed for an average male.[12] This is not only annoying but can even be dangerous in some aspects, one in particular is in drug development. For decades, medical research has focused on men as the standard human being. This has led to medicine being optimized for a typical white male, 70 kg, with way more muscles than a woman, and of course, without the pesky hormones. Men have a very different body composition than women, with around 18% of body fat compared to 25% in women. Women's kidneys also take much longer to clear substances from the blood compared to men. It is therefore very common that drugs stay longer in women's bodies, and if you keep filling up with more drugs at the same rate as men, the risk of getting a much higher dose of the drug than planned is very high. It is maybe important to note that the exclusion of women in clinical trials were not due to bad intentions but to a fear that clinical trials on women in the 70s had led to bad effects on foetuses.[13]

Drug-box

In an analysis of 163 new drug applications submitted to the U.S. Food and Drug Administration (FDA) between 1995 and 2000, drug concentrations in blood and tissues in 11 of the drugs varied by as much as 40% between men and women.[14] In antidepressants and antipsychotics, women tend to have higher drug concentrations in their blood than men do.[15]

The FDA has also recently been forced to change recommended doses in drugs such as the sleep aid Ambien, since the concentration stayed high in the blood too long. This led to the women being drowsy until late in the morning which could even lead to car accidents.[16] Women are also more likely than men to experience adverse reactions to drugs. Of all the drugs removed from the U.S. market between 1997 to 2001, 8 out of 10 caused more side effects in women.[17]

From 2016, the U.S. National Institute of Health requires that all new grant applications include both men and women, unless they provide a strong justification to exclusion of one sex.[18] That is a great step forward, but this is still just for government funded research. As most of the research is done directly in the pharmaceutical companies, the problem is not solved, and it will take time until it has changed.

Despite all that, it is an exciting time to be a woman. Many norms are being challenged, both in society and in science. Many discoveries in women's health have been made over the last decades. Some of them are astonishing, mainly because they were not known before. An example of such finding is the true size and shape of the clitoris, which was only discovered in the late 1990s. It was shown that what was seen of the clitoris was only the tip of the iceberg, and that in reality it is a large complex that surrounds both the vagina and the urethra.[19] It is a similar story with the anatomy of the breasts and breastfeeding, which got new explanations in 2005.[20] You will learn more about this in the pregnancy chapter.

It is a real concern that there is not enough research performed on women. It is, however, also important that the research that has been done is spread, implemented, and becomes common knowledge.

To really understand how the science works, let us start with the basics ...

THE MYSTERIOUS CHEMICALS

Hormones – the floating keys

Women often hear that they are 'hormonal', but what does that even mean? All humans, regardless of sex, are continuously influenced by hormones. They regulate most of our functions: our digestion and metabolism, how we breathe and sleep, if we get stressed, how we grow and develop, our mood, and of course how we reproduce.

Mostly, hormones act without us noticing anything, but sometimes we can feel the influence of some of them on our mood and on our wellbeing. As for me, getting 'hormonal' is synonym of being in high spirits and generally occurs when I have a high level of female sexual hormones, oestrogen. This is, however, not what most people refer to when they talk about being hormonal. They refer to this time of the month when women supposedly get cranky and difficult, just before their next period. In this phase of the menstrual cycle, all hormones that normally make us happy or calm are suddenly dropping from a high to a low level. Whereas the popular belief sees 'being hormonal' as being loaded with hormones, it is quite the opposite as we are rather on a hormonal detox.

Men also get hormonal, and this can sometimes be more problematic as it is less predictable and not cyclic in nature. There are links between aggressiveness and the male sexual hormone, testosterone.[21] We witness this link mainly in sports, but just as men are not victims of their hormones, neither are we. But what are these 'magical' hormones that have such an impact on our lives?

The word hormone comes from Greek and means 'setting in motion' and this is exactly what they do. You can picture the hormones as keys floating around in the blood, and whenever they encounter the appropriate keyhole, they will turn on the ignition and trigger a reaction. The triggered reactions occur in cells, both close to where the hormones are produced and far away in other parts of the body. If the nervous system is the hardwired communication system of the body, then hormones would be the Wi-Fi.[22] Hormones play a fundamental role in keeping the internal balance within the cells and within the whole body.

Most of the hormones are produced in glands, which together form the endocrine system. The endocrine glands in the body are situated in the brain (the hypothalamus, the pituitary gland and the pineal gland), the throat (the thyroid and parathyroid glands) and the abdomen (adrenal glands, the pancreas, the testes and the ovaries). Two of the most important glands which act as orchestrators for the other glands are the hypothalamus and the pituitary gland. They are situated in the brain and regulate how hormones are triggered in response to the world outside our bodies. The endocrine system oversees production of hormones and the regulation of how much of each hormone is released in the body.

The level of hormones in the blood depends on levels of other substances in the blood, like the amount of calcium, and by factors, such as stress, infections, and the content of fluid and minerals in the blood.[23]

Hormones can either act locally, or they can be released in the bloodstream, where they circulate until they reach a receptor cell ('keyhole'). When a hormone is captured by the receptor, the 'ignition' is turned on which triggers a reaction that changes how that cell behaves. By influencing millions of cells at the same time the hormones can induce large scale changes in the body.[24] The main purpose of hormones is to regulate bodily processes, not to control mind and behaviour, but they do, however, influence them in many ways.

Each person reacts very differently to the same amount of hormones. Indeed, the influence of a hormone does not only depend to its quantity, but also to the duration of its action, how many receptor cells are available, and to how well the receptors are working. The number of receptors can change from day to day, or even from minute to minute.[25]

The hormones are produced in many different types of cells, which make them sometimes appear at places they are not normally produced.[26] One example is oestrogen: in women oestrogen is mainly produced by cells in the ovaries, but it can also be produced in fat cells. As the hormonal world is complex, the fat cells produce different kinds of hormones depending on where they are located. If the fat cells are situated around a woman's waist, they will produce more male hormones, so called androgens, which can lead to hormonal imbalance and disruptions of the menstrual cycle.

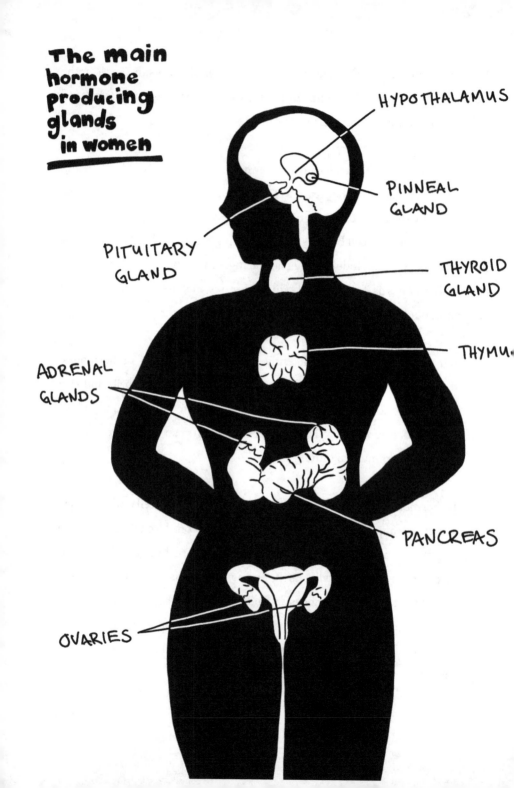

The main
hormone
producing
glands
in women

HYPOTHALAMUS

PINNEAL
GLAND

PITUITARY
GLAND

THYROID
GLAND

THYMU.

ADRENAL
GLANDS

PANCREAS

OVARIES

Understanding the action of hormones bring us closer to understanding the body but it still does not explain the full story. This body is way more complex than that. The truth is that we simply do not know or understand all the details. Most research in women looks at correlations between the amount of hormones and a certain phenomenon. But remember? Correlation is not synonym of causation and often elements will be missing in the story required to paint the full picture.

The hormonal axes

Hormones never act in isolation; they trigger chain reactions. Hormones released by glands in the brain will stimulate growth of cells in other parts of the body. As the cell grows, it will start producing another hormone which in turn will inhibit the hormones released by the glands. These types of interactions are called feedback loops, and the different glands affected by a certain feedback loop is called an axis.

Axis-box

One axis that is very relevant for women is the hypothalamic–pituitary–ovary axis (HPO axis) that coordinates the hormonal changes over the menstrual cycle.[27] Other important axes are the hypothalamic–pituitary–adrenal (HPA) axis, which regulates how we react to stress, and the hypothalamic–pituitary–thyroid (HPT) axis, which regulates metabolism and also to some extent stress. As you can see, the hypothalamus and pituitary glands are very important players in all these axes, which is one of the reasons why our reproduction and menstrual cycles are closely linked to our general wellbeing.

The endpoint of the HPO axis is the ovaries which produce the egg cells but they are also endocrine glands secreting the most important female sexual hormones, oestrogen and progesterone. The name oestrogen comes from the word 'oestrus' which means sexual desire and 'gene' meaning to generate. Generating sexual desire is by no means the only effect of oestrogen, but it does play a very important role in reproduction. It also has many other beneficial side effects like protecting hearts, brains and bones. A general effect of oestrogen, as you will see many examples of in the following chapters, is to help the cells grow more and create more blood vessels. This ability has many interesting side effects apart from their role in reproduction.

Progesterone means pro-gestation. It is, as the name indicates, the hormone to promote a pregnancy. It prepares the body for pregnancy by building up the wall of the uterus and making it ready for implantation of the embryo. After implantation, it maintains the pregnancy by preventing the uterus to contract and thereby expulsing the growing foetus. It has a general effect of relaxing smooth musculature and acts calming.

Testosterone on the other hand, simply means that it is a hormone produced in the testes, i.e., the testicles. This is obviously not the case in women, but the hormone was named before scientists discovered that women produce testosterone as well.

'Local hormones'

Apart from the 'classic' hormones scientists have recently discovered new 'chemicals' resembling hormones, eicosanoids. They act like hormones and trigger reactions in the cells, but on a local scale. This means that they mainly act on the cells close to where they are produced. However, just like normal hormones, they can sometimes get into the bloodstream and create actions in other parts of the body. There are several such local hormones, but the most studied and the type most relevant for reproductive health are the prostaglandins. The ingenious thing about prostaglandins is that they are not produced in a gland but directly in the cells close to where they are needed, triggered by chemicals released when a cell is injured. They induce healing by provoking fever, pain and inflammation.

In women, they play an important role in menstruation, and they initiate labour pains and ovulation. Its synthetic form is sometimes also given to women to induce labour.[28] Like other hormones, they act when they encounter receptors.[29]

Neurotransmitters

The hormones often work together with neurotransmitters. If the hormones are the Wi-Fi of the body, mainly using the bloodstream to get around, the neurotransmitters are a part of the hardwired system, the nervous system. Indeed, the latter promote the communication between the nerve cells and the target cells. Neurotransmitters are involved in nearly every function of the body, they are used to regulate heart rate, breathing, mood, sleep, concentration, appetite and muscle movements. Just like the hormones, neurotransmitters attach to different receptors where they trigger a reaction. There are more than a hundred different neurotransmitters. Some of them are well known, such as endorphins that give you the euphoric feeling after sports, dopamine responsible for giving you pleasure and reward after an activity – be it studying or gambling – or serotonin, the regulator of mood and appetite.

Now, when we have gotten a feel for the different types of hormones (but there are many more to discover), what purpose they serve and how they function, it is time to discover their effect on the female body. We will start with how they cyclically transform our body every month.

THE CYCLIC TRANSFORMATION

Over the course of approximately a month, the female body goes through a complete inner transformation. It is the cyclic preparation of our bodies to create new life, the menstrual cycle. When we think about the menstrual cycle, the bleeding is mainly what comes to mind, but the bleeding is just a sign that all the preparations went well and that our body has done its job. That includes developing and releasing an egg, setting up an environment to guide the sperm to the egg, and building up the wall of the uterus so, if fertilized, the egg can implant and start growing into a new life. If the egg is not fertilized, and hence you are not pregnant, all those preparations are torn down through the menstrual bleeding, and the preparation starts over again. All those changes happen on a monthly basis inside your body. They can be difficult to spot, but if you pay attention, you will notice that the hormonal changes that drive the transformation also influences many other aspects of life.

If you, as a woman, sometimes find yourself cursing evolution, demanding why you have to suffer through so many menstrual cycles, you are not putting your blame where it belongs. Having many menstrual cycles is an entirely new phenomenon. When the modern, average woman will have about 400 menstrual cycles between puberty and menopause and hence spend five to eight years bleeding, women in earlier times, and in societies where no contraception is used, would only menstruate around 100 times.[30] They spent much more of their time being either pregnant or breastfeeding and this has influenced our evolution. It is the new way of life during the last 60–70 years, with the capacity of controlling fertility, that has led to many more menstrual cycles. Exactly what the consequences will be is not yet clear but having so many menstrual cycles definitely change the overall hormonal exposure and increases the amount of blood lost over a lifetime.

Today, it is possible to suppress the menstrual cycles entirely using hormonal contraception. This brings the body in a state of fake pregnancy, which most likely also has consequences. We are just not sure how yet. Might it be beneficial because women were designed to be pregnant most of the time anyway? Or could it be harmful, as the artificial hormones are playing with the natural cycles? As with most things in medicine, the answer is not the same for everyone. But before exploring further the contraception topic, we need first to turn our attention to the natural menstrual cycle, a fine-tuned machinery if any.

Why cyclic?

One of the curious things about the menstrual cycle is that it is just that, a cycle. Why are we not continuously prepared for pregnancy like most other species? Rabbits, for instance, ovulate spontaneously when they have sex, and other animals only build up the lining of the uterus once they are already pregnant. This means that they are not shedding it monthly in a bleeding like humans do. The only other menstruating animals we know of are other primates, bats and elephant shrews.[31] (In case you think that elephant shrews are female elephants, it is not quite that. Elephant shrews are small mammals resembling rodents, and have very little to do with real elephants, apart from a trunk-like nose.) Some other animals do bleed on occasion, as owners of dogs and cats might have noticed, but that bleeding comes from the vagina and not the uterus.

It is difficult to understand why human fertility is cyclic, and there are many theories to explain it. Currently the one rallying the most scientific voices asserts that women, as mothers, can unconsciously control what is happening in their own womb.[32] Indeed we only ovulate if our body is fit enough for pregnancy. If we are in a stressful situation, like if we are not well nourished, ovulation will be suppressed and the progesterone, which together with oestrogen is crucial to build up the lining of the uterus, will not be triggered. Hence, we will not be able get pregnant. In this way, the body is deciding if pregnancy is a good thing at this moment or not.

The current theory is that in humans, the uterine lining needs to be there before the egg and the embryo implants because a human embryo is more demanding than many other species. A human foetus needs more nutriments and hence digs deeper into the lining to be able to profit from the mother's blood. In animals, the embryos stay more on the surface of the uterine lining. This new uterus wall, built before the egg implants, protects the mother from the hungry foetus and that would explain why the uterine wall is in place already before pregnancy.

The cyclic strategy could also be explained as a way of getting rid of bad embryos. Compared to other species, human embryos are very prone to genetic abnormalities. This is why there is so many early miscarriages. It is believed that up to 50% of all pregnancies end in miscarriage, but most of them happen before you know you are pregnant. After a pregnancy has been detected, that number is still 15–25%. This might sound harsh, but it is an ingenious construction; the cells of the uterine lining are capable of recognizing embryos with genetic defects and can therefore shed the uterine lining and get rid of the embryo i.e., the body doesn't need to waste resources on a pregnancy that might not be successful later on.[33]

Another theory claims that getting rid of the uterine lining on a monthly basis saves energy. Remember that evolution happened before we worried about eating too many calories and having enough of them was the major concern. Having a fully built uterine lining requires 7% more energy compared to not having one. By shedding it every month, up to 6 days' worth of food can be saved over four menstrual cycles.[34]

Understanding the cycle

There are many reasons why it is useful to understand how the menstrual cycle works apart from knowing when you need to pack pads in your purse. The most obvious one has to do with fertility. If you want to get pregnant without medical intervention, a healthy cycle is the minimum prerequisite. By knowing the cycle, we can also prevent pregnancy by abstaining from sex on the fertile days, but there are more benefits. A healthy menstrual cycle can be used as a sign of the overall health. For girls and teenagers, both the American Academy of Pediatrics (AAP) and American College of Obstetricians and Gynecologists (ACOG) recommend using the menstrual cycle as a vital sign, next to pulse rate, temperature and other, to assess normal development and detect diseases.[35] Having a healthy cycle bring many long-term benefits and protection from conditions such as heart disease, dementia, osteoporosis and others.

What is it and how does it work?

The menstrual cycle starts on the first day of the bleeding period and lasts until the day before the next bleeding. This makes it very easy to recognize the beginning and the end, but trickier to recognize what happens in between. The ovulation (the moment where the ovary releases an egg) happens somewhere in the middle of the cycle. The 3 to 7 days leading up to and including ovulation is the fertile phase. It is only during those days that one can get pregnant, or rather, it is only during these days that the sperm can survive sufficiently long inside and get a chance to fuse with the egg. The phase before ovulation is called the follicular phase, and the period after the luteal phase.

The menstrual cycle is an amazing coordination act between many different hormones. One hormone stimulates the growth of certain cells which start producing other hormones. Thereafter, those new hormones start acting as inhibitors in order to turn off the first hormone, and so it goes, over and over, driving the cyclic nature of the menstrual cycle.

The simplified version of the menstrual cycle

Very simplified, in the beginning of the cycle, follicle stimulating hormone (FSH) turns the follicles, waiting in the ovaries, into eggs. This is why the first half of the menstrual cycle is referred to as the follicular phase. The body prepares several follicles each cycle but only one will be selected to become a fully developed egg.

In the second half of the follicular phase, about a quarter into the cycle, oestrogen level increases. This stimulates the cervical mucus to create a nice and comfy environment in the vagina, uterus and fallopian tubes for the sperm to swim and live while waiting for the egg to arrive. This period is called the fertile phase or the fertile window, and it is only during this period that having intercourse will lead to pregnancy. In this phase, many women are more aroused, which is of course a nice little trick of nature to keep us reproducing. There are also many indications that we become more attractive in this phase. Studies have shown that lap dancers working in their fertile phase receive more tips than their colleagues.[36]

After the oestrogen has started to rise, there will be a strong and short peak of the luteinizing hormone (LH), which triggers ovulation. During ovulation, the egg is swept from the ovaries into the fallopian tubes, where it will, or will not, meet the waiting sperm.

Around ovulation, progesterone is triggered. The purpose of progesterone is to build up the wall of the uterus so that, if fertilized, the egg can implant and start growing.

If the fertilized egg implants, meaning that you are pregnant, the progesterone level will keep rising. If not, both progesterone and oestrogen will fall back to the baseline level, and the wall of the uterus will be torn down in the form of bleeding.

The short description above covers the basics, but there are many interesting details about what is happening in the body: the hormonal interplay and the complete transformation of the reproductive organs.

The main players of the hormonal cycle

I realize that not everyone gets crazy enthusiastic about reading the nitty gritty details of the menstrual cycle. If you are one of those, you should not feel bad to jump directly to the subchapter about ovulation on page 52, but for my fellow geeks, I recommend you keep reading.

The hormonal coordination of the cycle has three main players, namely the hypothalamus, the pituitary gland and the ovaries.

The hypothalamus – the orchestrator

The hypothalamus is a small, almond-sized, structure in the brain with many different functions. One of its main responsibilities is the coordination between the nervous system and the endocrine system. It is a part of the limbic system which coordinates many of our basic instincts like emotions, behaviour, motivation and memory.

Among others, the hypothalamus produces a hormone with a complicated name: gonadotropin-releasing hormone (GnRH). Gonadotropin is not the name of a fertility goddess but a combination of medical terms; the suffix '– tropin' means that it has a stimulating effect on something, so gonadotropin means that it has a stimulating effect on the gonads. If you remember from the introduction, the gonads were the glands we have when we are embryos. They turn into testicles in men and ovaries in women while we are still in the womb. In women, gonadotropin regulates normal growth, sexual development and reproductive function. As mentioned above, the hypothalamus does not re-lease gonadotropin but GnRH, meaning a hormone that tells another part of the body to get to work and start producing gonadotropin. That other part is

THE
HPO-AXIS

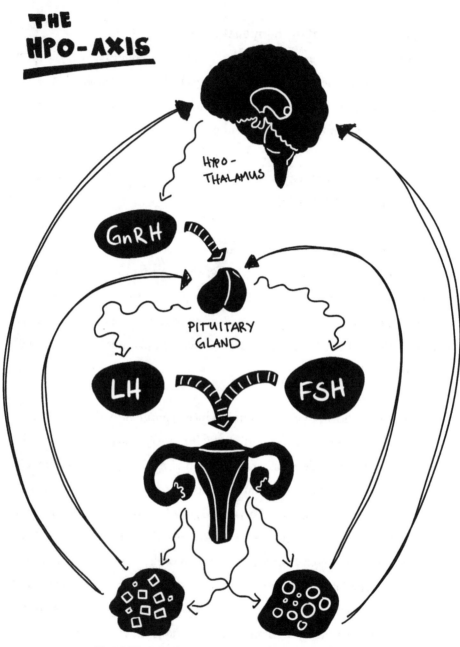

the pituitary gland, or more precisely, the anterior pituitary gland which is a subpart of the former.

The anterior pituitary gland – small but powerful

The anterior pituitary gland is only the size of a pea, but has an importance largely exceeding its modest dimensions. It is situated just below the hypothalamus and is also known as the hypophysis. Its main role is to synthesize and secrete hormones into the blood. Apart from reproduction, this little structure is responsible for regulating many important processes such as stress, growth, and breastfeeding.

When the anterior pituitary gland receives the GnRH from the hypothalamus through a direct vein, it starts producing gonadotropins. The gonadotropins are a group of hormones. Two of them are important for reproduction: the follicle-stimulating hormone (FSH) and luteinizing hormone (LH), which are both released into the bloodstream through which they will reach the ovaries.

The ovaries – storage of the future

The ovary is the reproductive warehouse of the body. Here the follicles are stored and will get transformed into eggs, before being transferred into the fallopian tubes. One egg per cycle. There are two ovaries, one on each side of the uterus; they are connected to the uterus through the fallopian tubes. These two warehouses are already filled when we are born; all the follicles we will ever have are already in place in the ovaries while we are in the womb. They are formed when the mothers are only 6 weeks pregnant, and we are just a tiny foetus; by the time she is 20 weeks pregnant we have 6 to 7 million of them. Thereafter, they start decreasing in number; by the time we are born there are about 2 million follicles left and at puberty around 30,000. We still do not know why nature thinks such a terrible waste of follicles is a good idea, and the wasting will continue throughout the entire reproductive age. Only about 400 of the follicles will eventually become eggs and be released through ovulation between puberty and menopause. By menopause, the storage of follicles is strongly diminished, but the main problem is that they lose their capacity to grow and lead to ovulation.

Yellow-box

When the gonadotropins released from the anterior pituitary gland reach the ovaries, they will attach to receptors there and start acting. As its name suggests, follicle stimulating hormone (FSH) will stimulate the follicles to grow. The follicles will produce oestrogens that lowers FSH through negative feedback. Only the largest follicle will survive this drop in FSH. When the dominant follicle is large enough, it will produce high levels of oestrogens triggering a surge of luteinizing hormone (LH). The LH will make the dominant follicle release the egg and will develop what is left of the follicle into a cellular, hormone-producing mass, called the corpus luteum. 'Luteum' means 'yellow' in Latin, so the corpus luteum refers to 'the yellow body'. The role of the corpus luteum is to produce progesterone, which has the very important job of preparing the wall of the uterus to receive the fertilized egg. Progesterone is also essential in maintaining a pregnancy, hence the name 'pro-gestation'.[37] As the corpus luteum will only be formed after a successful ovulation, progesterone is a sign that ovulation happened.[38] In addition to progesterone, the corpus luteum also produces oestrogen but in a lower amount.

Progesterone and oestrogen will inhibit the production of LH and FSH, which is no longer needed in this phase. The previous surge of LH will stimulate the corpus luteum for about a week but after that, the only thing that can save the short life of the corpus luteum is that the egg is fertilized and implanted in the uterus, i.e., if you are pregnant.

The hormones that are produced by the ovaries have an immediate effect on their close surrounding, like fallopian tubes and the uterus, but as they are released into the bloodstream, they also impact many other parts of the body. Once they return to the anterior pituitary gland and the hypothalamus, they start inhibiting the production of gonadotropins and gonadotropin-releasing hormones.[39]

Scientist-box

A pioneer of progesterone research and explorer of the full function of the corpus luteum was the German Ludwig Fraenkel (1870–1951). In the beginning of the twentieth century, he proved that the corpus luteum was an endocrine organ and he was the first to purify progesterone.[40] He was very active in the gynaecological societies during the first decade of 1900s and demonstrated that social living conditions impact on gynaecological illnesses. As a Jew, he was forced to leave the country after the Nazis had taken the power in Germany; he moved to Uruguay.[41]

The intricate play of the hormones

The hypothalamus, the anterior pituitary gland and the ovaries constitute a communication axis named the hypothalamus–pituitary–ovary axis (HPO-axis). It is the intricate play of hormones along this axis that forms the menstrual cycle. The hormones produced by one of these players will either stimulate or inhibit production of hormones in the other parts.

The menstrual cycle
Simplified

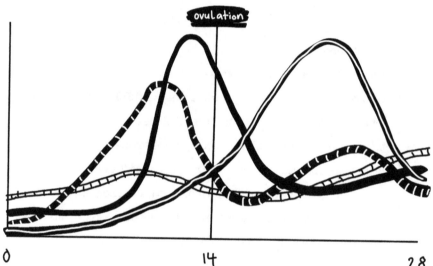

- LH (Luteinizing hormone)
- E2 (Oestrogen / Estradiol)
- FSH (Follicle stimulating hormone)
- PG (Progesterone)

ovulation

0   14   28

neutral

happy!

horny

calm

cranky

The two main players situated in the brain, the hypothalamus and the anterior pituitary gland, are also involved in many other important processes in the body such as stress and emotions. It is therefore not so surprising that our surroundings and wellbeing have an impact on our menstrual cycle[42] and that the menstrual cycle has an impact on our mood and wellbeing. The hormones released from the hypothalamus and pituitary gland prepare the egg to be released from the ovaries and hence play their main role in the follicular phase. Consequently, the follicular phase is the most sensitive to disturbances such as stress, hunger, and other hormones, and it is therefore this phase has the highest variation in length.

Detailed-box

We are mainly talking about oestrogen, but oestrogens are a group of similar hormones which includes oestradiol, estrone and estriol. The most active of the three – and the one that is produced in highest abundance in the ovaries – is oestradiol, and it is the one I am generally referring to when I write oestrogen. It is interesting to note that the oestrogen produced in the follicle is synthesized from androgens, the hormones that were traditionally considered male. The presence of FSH is necessary to stimulate this chemical reaction.

Another hormone that is also produced inside the growing follicles is anti-Müllerian hormone (AMH). AMH is needed for oestrogen production and the amount of AMH is related to the number of recruited follicles. The number of recruited follicles is dependent on the overall ovarian reserve, which is why measuring AMH is a good indicator of how many years are left to menopause[43] but not necessarily a good measure of how likely you are to get pregnant.

Other hormones released from the ovaries together with the two already mentioned are different types of inhibin. The work of inhibin is, as you might expect, to inhibit production of hormones elsewhere.

Ovulation – the gentle act of letting go

During the follicular phase, i.e., the first half of the menstrual cycle, the follicle is preparing the egg (or the oocyte which is the medical term) to be released while transforming the follicle itself. The wall of the follicle becomes thin and stretched just before ovulation,[44] and the ovulation being the very moment when the follicle opens to release the egg, or the oocyte. It is not a 'pop' but a smooth process. The whole process is stimulated by prostaglandins, a hormone weakening the wall of the follicle. Prostaglandins are vasodilators, they weaken the wall of the follicle and play a role in the inflammation process. Therefore, anti-inflammatory drugs such as ibuprofen can inhibit ovulation by preventing inflammation, the prostaglandin effect.

Surprisingly, the ovaries are not connected to the fallopian tubes, so it is not quite clear how the egg actually reaches the fallopian tube. One theory is that the small, finger-like ends of the fallopian tubes sweeps over the ovary, catches the egg, and draws it into the tube through muscular control. However, this theory is challenged by the fact that women have gotten pregnant while ovulating on one side where there is an ovary but no tube.[45]

Timing of ovulation

If you are trying to get pregnant using 'timed intercourse' – meaning that you are trying to have sex during your fertile days to maximize your chances to get pregnant – you are probably acquainted with LH tests. These tests measure the amount of LH in the urine. Ovulation takes place about 12 hours after LH has reached its peak and 24–36 hours after the highest point of oestrogen. It is difficult to pinpoint precisely at what moment the LH has reached the maximum, as a maximum can only be seen after the hormone levels start going down again. At this moment it is often too late to have intercourse to get pregnant. Therefore, the most useful way of estimating ovulation to identify the fertile phase is to look for the moment when LH starts to rise. This happens around 34–36 hours before ovulation.[46]

There are many different LH tests available on the market. Some of them measure the amount of LH and leave it to the user to estimate when the rise occurs, which can sometimes be tricky. In the clinical trials I have been involved in, we found it easier to get a precise estimation of ovulation day using digital tests that do that estimation for the user. The LH tests are the most reliable way of pinpointing the exact timing of ovulation without the assistance of an ultrasound. However, even those tests allow you very little time to act, since you need to have sex before or within hours of the ovulation. Hence, some women – and quite a few men as well – find it stressful and prefer to use other methods to track the entire fertile window. There are also indications that using LH tests in general might lead to being too late, both in timing intercourse and for IVF interventions. A better method to find the most fertile period would be to track the cervical mucus.[47] This is basically the body's own way of showing when it is fertile.

Thanks to all the work that has been done in improving IVF treatments (in vitro fertilization as a way of helping couples get pregnant), a great amount of knowledge on the timing of ovulation has been accumulated over the last decades. In two-thirds of all women, the LH surge happens between 3 am and 8 am.[48] It also seems like the exact timing of ovulation depends on what time of the year it is. During spring, ovulation mainly happens in the morning and during the autumn and winter mainly in the evening. In the northern hemisphere, it has been proven that 90% of all women ovulate between 4 pm and 7 pm between July and February (12 hours after the LH peak).

For unknown reasons, 55% of ovulations happen in the right ovary and even though it does not seem to influence the cycle characteristics, there seems to be a higher potential for pregnancy when you ovulate from the right ovary.[49] The right ovary is also known to be slightly bigger than the left, which is also the case for testicles, with the right one often being larger than the left. There is a popular belief that penises tend to pivot more towards the left. Unsurprisingly, there are few clinical studies on that topic, so I am not able to confirm that claim, but having a heavier right testicle could of course contribute to that fact through pure mechanics.

Assisted conception – helping nature

During an IVF procedure, the eggs are taken directly from the follicles inside the ovaries just before the follicles are ready to release them. The women who go through IVF need to take additional hormones (FSH) for the follicles to produce more than just one dominant follicle, so more than one egg can be retrieved at a time.[50] Towards the end of the follicular phase, the woman receives a shot of another hormone, usually the human chorionic gonadotropin (HCG) that is also produced during pregnancy. This hormone has a similar effect as the LH one, which is the stimulation of the final maturation of the oocyte and its release from the wall of the follicle. The shot will trigger ovulation about 36 hours later: around that time, the doctors will intervene to retrieve the eggs.[51]

Women going through a hormonal IVF treatment, either taking FSH or clomiphene to boost ovulation, tend to have several follicles evolving, which makes twin births more common. When several eggs are released and fertilized, we talk about fraternal twinning, which is not the same as identical twins (who come from the same egg). Fraternal twins are as similar as any siblings and can for instance be a boy and a girl.[52] Women approaching menopause have higher levels of FSH, which is why twin births are more frequent in older women. When more than one egg is released, it can either be from the same ovary or from both.

Twenty years ago, already, scientists were able to remove a part of the ovarian tissue, freeze it, and implant it back again in order to use the follicles to develop into new eggs and to produce oestradiol.[53] In Denmark, this procedure is used routinely by some clinical teams to help young women maintain their fertility after a cancer treatment, which would normally lead to infertility. The outcomes are very positive with the ovary tissue being active up to ten years after the implantation and about a third of the women managed to get pregnant, of which around half were natural conceptions.[54] In the UK, the very same procedure is also offered to delay menopause by implanting ovarian tissues from your younger self in order to keep producing eggs and hormones.[55] It is, however, still an experimental procedure and is not recommended as a method to delay childbirth.[56]

As for egg freezing, the success rate for giving birth using your own eggs was only 18% in the UK according to a report published in 2018. Using frozen eggs from another donor increased the chances to 30%. The reason behind this difference is that women often decide to freeze their eggs late in their reproductive life; most women were 38 years and older. With eggs from donors that were 35 and younger, the success rates increase considerably.[57]

Ovulation pain (mittelschmerz)

When I tried to get pregnant, I had just gotten off the pill, so I did not know my cycle very well. I therefore assumed that I would be like the average woman and ovulate around day 15. About a week after my presumed ovulation, I would always feel a pain in my lower abdomen, which I would interpret as the first signs of pregnancy. As it turned out, however, what I was feeling was not pregnancy but ovulation pain. I was ovulating much later than I thought. The majority of women cannot feel when they are ovulating and for those who can, the level of pain varies a lot. Normally, the pain only comes from one side of the abdomen depending on which ovary you are ovulating from. The medical term for this ovulation pain is *mittelschmerz*.[58] It is a German term which means 'pain in the middle'; the middle not referring to the body but to the phase of the menstrual cycle. What exactly causes the pain is not known, but when the egg breaks through the wall of the follicle, ovary fluid and some blood are released into the abdomen, which might irritate the surrounding nerves.

The cyclic transformation of the reproductive organs

If the purpose of the follicular phase is to prepare the body for releasing a fully functional egg, the role of the luteal phase is to prepare the physical environment for pregnancy. That covers everything from bringing the egg safely from the ovary to the uterus, helping the sperm find and fertilize the egg, building up the wall of the uterus for the fertilized egg to implant and nurture the fertilized egg so it can start growing. It is common to see the sperm as the brave swimmer that conquers all obstacles to reach the egg, but the truth is that the female reproductive system transforms its entire environment to select the best sperm and then accommodate and guide the selected sperm through the whole journey.

Fallopian tubes – the true centre of action

The journey of the egg starts in the fallopian tubes, where the egg is swept in immediately after ovulation. The word 'swept' is carefully chosen as the ends of the fallopian tubes have a set of 'sticky fingers' that sweep over the surroundings and, to our present knowledge, this is how the egg gets taken from the ovary and transferred into the tube. The fallopian tube is the channel that connects the ovary to the uterus, but it does more than simply transporting the egg. Indeed, it assists in every way it can the encounter between the egg and the sperm and prepares both for fertilization. The fallopian tubes also harbour the early embryo and helps it develop, while the latter is waiting to continue to the uterus.

The egg and the sperm enter the fallopian tubes from opposite sides. The sperm makes its entrance to the fallopian tube from the uterus, and the egg is swept in there from the ovaries. The environment in the tube, including both fluid and muscle movements, is influenced by the presence of sperm in the tract and is modified to do everything to help and guide the sperm to encounter the egg in the middle. While looking for the egg, the female body has prepared the sperm through a chemical process called capacitation. The result of this is hyperactivation: i.e., the swimming patterns of the sperm change – accelerating and intensifying – and its tail slaps faster. These transformations enable the sperm to penetrate the egg.[59]

After fertilization, the embryo stays in the fallopian tube for 2 to 3 days, where it is well nurtured. By keeping the embryo in the fallopian tube for a while before transporting it into the uterus, the latter has time to build up its wall and be ready to welcome the embryo. While developing, the early embryo is slowly transported to the uterus by a combination of movements of cilia, muscle con-

Journey of a fertilized egg cell

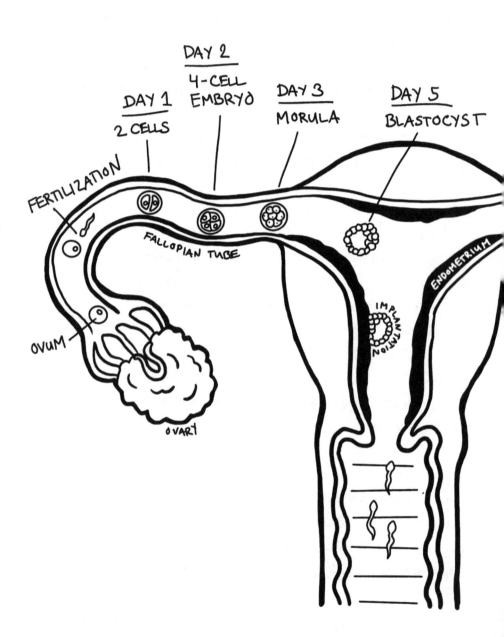

tractions and fluid flow.[60] It will take 3 to 4 days after ovulation for the egg to complete its journey through the fallopian tube and reach its destination where it can implant and start growing.

Fallopian-box

Having blocked fallopian tubes is a common cause of infertility, but it is very difficult to detect. A friend of mine who had trouble getting pregnant got her tubes flushed with liquid, which is a procedure known for decades but has recently undergone some improvements with the use of oil-based fluid.[61]

If it had anything to do with the flushed tubes or not remains unclear, but my friend got pregnant shortly after the procedure. In female sterilization, the fallopian tubes are blocked with a small clip or a ring.[62] Problems with the fallopian tubes can also have more serious consequences, for they are the most common site of ectopic pregnancies.[63] An ectopic pregnancy occurs when the embryo implants outside the uterus which can be very dangerous for the mother and cause severe internal bleeding. Even though modern medicine has made tremendous progress, it is still an important cause of mortality in pregnancy in countries with difficult access to high quality healthcare. Sadly, in the current state of medicine the foetus can't be saved in case of an ectopic pregnancy.

The uterus – "The palace of seeds"

The ancient Greeks were terribly obsessed with women's uteruses, or wombs. Both the physical and mental differences between men and women were believed to originate from a condition called the 'wandering womb'. It was believed that it was moving around linked to different illnesses women might have and especially to mental issues such as 'hysteria'. The physician Aretaeus of Cappadocia went so far as to consider the womb "an animal within an animal."[64]

The Chinese have a more positive way of describing the uterus. Their sign for womb is the combination of the signs for palace and seed.

子宮

Zĭgōng

Preparing for implantation

The uterus might not be an animal nor has the power to transform us, but it is an incredible structure. The uterus is the place where life is literally created, where the whole development from embryo to a fully developed baby takes place. The most extreme physical transformation in an adult woman, apart from pregnancy itself, is the cyclic preparation of the womb to receive the fertilized egg. The wall of the uterus is covered with a cell layer called the endometrium which might be one of the most complex tissues in the human body, as its continuous changes bear witness.[65] At every menstrual cycle the endometrium is rebuilt to host a life, just to be torn down if no fertilized egg arrives. It is even capable of evaluating the quality of the fertilized egg, and selectively deciding whether to allow it to implant in the uterus.

Just after menstruation, the endometrium is at its thinnest and only consists of a fine layer of cells. Under the influence of oestrogen, produced during the follicular phase, the endometrium starts to reconstruct and grow again by building new cells. As you approach ovulation, the endometrium keeps thickening with more blood vessels. The endometrial glands are normally small but after impregnation they become enlarged.[66] The role of the endometrial glands – especially those in the region of the cervix (the gateway between the uterus and the vagina) – is to produce cervical mucus, which will help to guide the sperm towards the right destination.

Gooey fact

Cervical mucus is a gel-like fluid that is secreted through the vagina and changes in thickness and amount with the fluctuating hormones. The fluid produced in this phase with high oestrogen level has a special egg-white consistency. You can feel it with your fingers in the vagina and with some practice, it can be used as a sign that you are in your fertile phase.[67] The cervical mucus also serves as a protection for the uterus and thanks to its antibacterial properties it can prevent some infections.[68] The cervical mucus forms the vaginal discharge that you sometimes find in your underwear. It protects your vagina from infections and keeps it clean and moist.

The cervix itself also goes through changes during the menstrual cycle. It is normally firm, positioned low in the vagina, and closed. Just before ovulation, with the influence of oestrogen, it becomes softer, moves higher up, and opens.[69]

Under the influence of the oestrogen and progesterone produced by the corpus luteum, the endometrium keeps building up, getting stronger and stronger. At the time of ovulation, it is 3 to 5 millimetres thick, but it will keep growing. It needs to be at least 6 mm thick to be able to receive an implantation.[70] The oestrogen helps the cells of the endometrium to grow and multiply for a few more days. The progesterone makes the endometrium swell and produce a highly nutritive 'uterine milk', which nourishes the fertilized egg from the moment it reaches the uterus from the fallopian tube until it implants in the uterine wall about 7 to 9 days after ovulation.[71] Once the egg is implanted in the endometrium, it starts getting its nutrients directly from the endometrium.[72]

The cyclic transformation of the utcrus

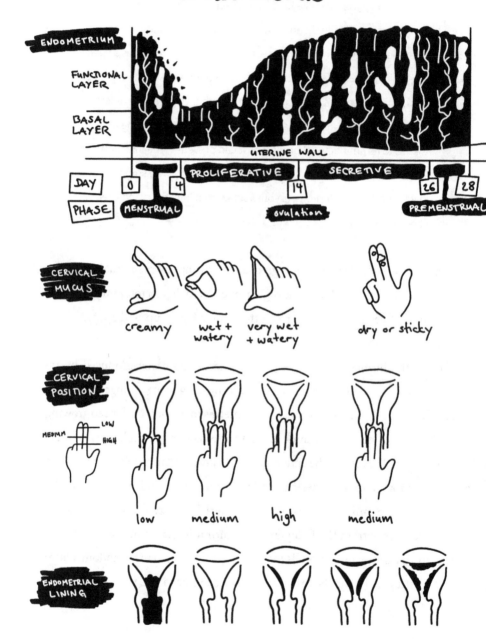

Preparing to shed and menstrual pain

When we are not pregnant it means that the egg was not fertilized, or the fertilized egg did not implant in the endometrium. Without the pregnancy hormone (hCG), the corpus luteum is destined to shrink and disappear. The degradation starts about a week after ovulation and leads to a strong decrease in both oestrogen and progesterone. The drop leads to many premenstrual symptoms. In the uterus, the dip in progesterone leads to an increased production of prostaglandins in the cells of the endometrium. Prostaglandins – signalling molecules that resemble hormones – play an important role in inflammation. This is the reason why we say that menstruation is an inflammatory event. The prostaglandins cause contractions of the uterus while cutting off the blood vessels.[73] The strangling of the blood supply minimizes the blood loss during menstruation and reduces the flow of nutrients to the endometrium.

Period pain occurs when the contractions of the uterus are very strong and cuts off the oxygen supply to the tissue. The strength of the contraction is linked to the level of prostaglandins, and it has been shown that women with severe menstrual pain often have an excess of prostaglandins, which contract uterus.[74] As ibuprofen and other nonsteroidal anti-inflammatory drugs (NSAIDs) work actively against prostaglandins, taking them can relieve both the pain and the amount of menstrual flow.[75]

The uterus does not spend much time mourning over lost tissue, but immediately starts rebuilding the endometrium. This takes place in some areas of the uterus, whereas others are still being shed.[76] Shedding the endometrium little by little minimizes the area of exposed and bleeding surface, which

Menstruation

IF THE EGG CELL IS NOT FERTILIZED, IT IS SHED ALONG WITH THE **ENDOMETRIUM**

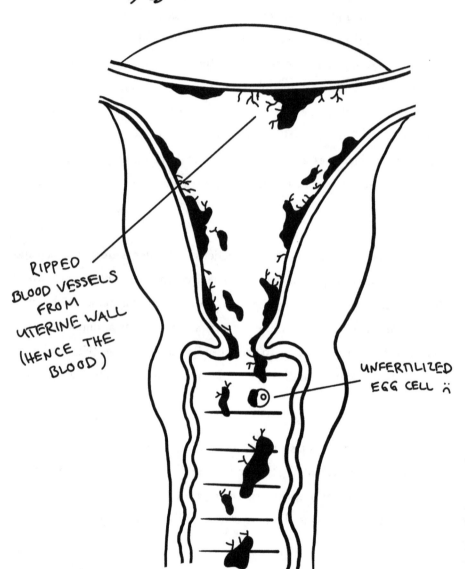

RIPPED BLOOD VESSELS FROM UTERINE WALL (HENCE THE BLOOD)

UNFERTILIZED EGG CELL ä

minimizes in turn the risk of infection. Once the endometrium is healed and rebuilt – which normally occurs within 4 to 7 days after the bleeding started – the bleeding stops again.

As the uterus is not a part of the hormonal interplay, it is possible for women to keep their hormonal cycles even after the uterus has been removed through hysterectomy.[77]

Endometriosis – menstruating on the inside

In the brilliant novel *Conversations with Friends* by Sally Rooney, the main character Frances develops period pains so paralyzing that she faints and thinks she is going to die. Unfortunately, this is not a purely fictional description of how many women feel during the first days of their menstrual bleeding. Severe period pain is, however, not normal and in the case of Frances, it was due to a condition called endometriosis.

In endometriosis, endometrial cells from the uterus have leaked out of the uterus and fallopian tubes into the pelvis and start growing outside of their normal environment. Whereas the endometrium inside the uterus is crucial to help the foetus implant, endometrial cells outside of the uterus can be very harmful. The endometrial cells outside the uterus follow the same hormonal influence as the endometrium on the inside: they multiply and grow with oestrogen, swell with progesterone, and are shed during menstrual bleeding. This causes bleeding and prostaglandin production inside the pelvis, which leads to inflammation and severe pain. The pain can be more or less serious depending on how close the endometrial cells are to nerve endings.[78]

Endometriosis

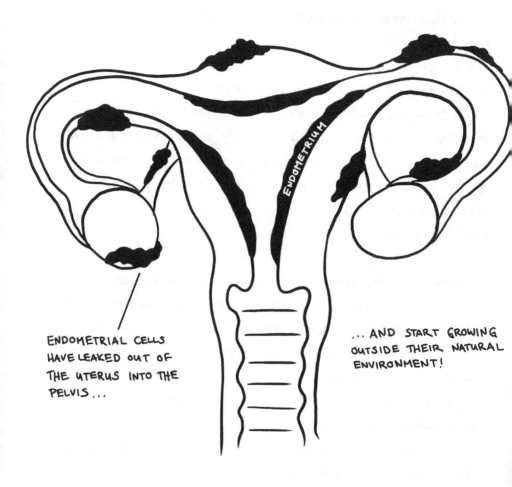

ENDOMETRIAL CELLS HAVE LEAKED OUT OF THE UTERUS INTO THE PELVIS...

... AND START GROWING OUTSIDE THEIR NATURAL ENVIRONMENT!

But how is it possible for the endometrial cells to leak from the uterus into the pelvis? Most likely, it is due to backwards menstruation: instead of being transported out of the uterus into the vagina and out of the body, the menstrual fluid is going back into the fallopian tubes. There it can leak out into the pelvis and the endometrial cells that normally should be flushed out with the menstruation, can implant on the outside of the fallopian tubes and the uterus. The endometrial cells then stay there and keep following the hormonal cycle. In rare cases, the endometrial cells can even travel to other parts of the body with the bloodstream and appear on other organs such as the bladder, bowel, lungs, and liver. Depending on where the cells end up, you can develop different types of endometriosis, the most common one is bladder endometriosis which might lead to aches, painful urination and other symptoms.[80]

In general, the more you bleed, the higher the risk for endometriosis. There are many reasons for excessive bleeding; it could be if you had your first period early, or if you have a late menopause. It could also be that you have never been pregnant, that you have short menstrual cycles or have heavy periods with bleeding longer than 7 days. You also have a higher risk if you have a low BMI or a family history of endometriosis.[81] The first symptoms can appear at any age.

Endometriosis impacts fertility in many ways. If the endometrial cells are implanted in a manner that enables them to 'eat' into the ovaries, they can influence the number of eggs. On the fallopian tubes, the cells can create scars, which can clog the tubes and prevent the eggs' passage and meeting with the sperm. As endometriosis creates inflammation, it can both inflame the uterus, prevent the implantation of the egg, modify the environment, and reduce the sperms' capacity to swim,[82] thereby lowering the chance of pregnancy or altogether preventing it.

Endometriosis is estimated to be the reason behind up to 50% of all infertility; globally, 6–10% of women suffer from endometriosis. Since no test exists to detect it, it is very difficult to diagnose it. To this day, the best way to confirm endometriosis is through ultrasound or magnetic resonance imaging. On average a woman needs to see seven physicians before being diagnosed.[83]

The most common treatment of endometriosis is to suppress menstruation by using the pill (without hormone free days) or by using a hormone releasing intrauterine device which releases synthetic progesterone.

Uterine fibroids and adenomyosis

When she was around 30 years old, a friend of mine started getting stronger and stronger menstrual bleedings to a point where she would have to report sick to work almost every month. If she managed to show up anyway, she would put in a pad before leaving home that needed to be changed immediately when she arrived. The bleeding was such that she would become anaemic and would need an injection of iron before and after her menstrual bleeding. Not only was the pain excruciating – she described it as if someone were trying to pull out her uterus, and her kidneys too would hurt – but she would also feel exhausted and swollen just before her periods.

Fearful that it would be a lethal cancer, it took her a year and a half until she dared to go see her gynaecologist. After an MRI scan and a biopsy, they discovered that she had a very large uterine fibroid. Uterine fibroids are benign (not dangerous) tumours growing inside the uterus. They can vary in size but are normally not large enough to be seen by the human eye. However, in her case, she had a huge one weighing 2 kg. In order to get it out, the doctors needed to

Adenomyosis

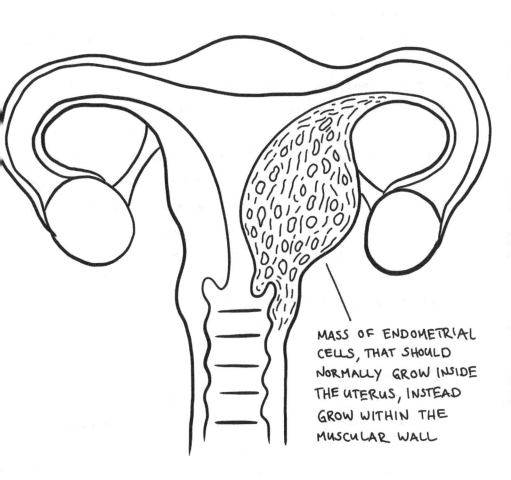

MASS OF ENDOMETRIAL
CELLS, THAT SHOULD
NORMALLY GROW INSIDE
THE UTERUS, INSTEAD
GROW WITHIN THE
MUSCULAR WALL

remove her entire uterus. My friend is healthy again but is living with the grief of not being able to have children.

My friend's case is a rare one. Most women with fibroids show no symptoms, but if they do, they include long and heavy menstrual periods, period pain, pelvic pain, constipation and frequent urination.[84] Having smaller fibroids is very common and, by the age of 50, it is estimated that as many as 80% of all women have them to some extent.[85] African-American women are highly affected by this ailment and also suffer from stronger symptoms.[86]

Uterine fibroids and adenomyosis are often mixed up because they have very similar symptoms but whereas fibroids are small, benign, well-defined tumours, adenomyosis is a less defined mass of cells growing inside the muscular wall of the uterus. Those cells are the endometrial cells that should normally grow inside the uterus but that have, as in the case of endometriosis, escaped their natural environment. Adenomyosis is not dangerous but it can lead to heavy periods, strong period pain and sometimes even chronic pelvic pain.[87]

The bloody details

As we have seen, many extraordinary things happen between periods, but we do spend a significant time bleeding. From our first period around 12 years of age until the last at around 50, we bleed 20–25% of the time. Whereas most mammals signal their fertility in a very visible way, such as the red, swollen buttocks of the fertile baboon, humans handle their fertility more discreetly. The only visible indication that we get of our fertility is the monthly bleeding.

One of the old theories about menstruation is that it was a way to get rid of toxins and bad sperm, but that idea has since long been debunked. We talk about blood, but the menstrual fluid is much more than that: almost half of it is made of endometrium and other fluids. Normally, free floating blood will clot, but together with the endometrium there is also a substance called fibrinolysin, which is released to prevent the blood from coagulating. However, if large amounts of blood are shed, the available fibrinolysin might not be enough to prevent clotting. Hence, the presence of blood clots can be an indication of an underlying issue.[88] There is no reason to worry about a small amount of blood clots, but if it becomes the norm, it could be worth checking. According to Healthline, normal blood clots should be smaller than a quarter, not be too frequent and occur at the beginning of the cycle.[89] Ah, and for all non-Americans, a quarter is a coin about 2.5 cm in diameter.

A large number of white blood cells (leukocytes) are also released with the blood. White blood cells are the cells of the immune system that are involved in protecting the body against both infectious disease and foreign invaders. Because of the release of these leukocytes, the uterus is very resistant to infections during menses even though the endometrium is shedding.

There are no medical reasons to not have sex during menses but it is important to remember that the blood does not protect you against sexually transmitted diseases or infections, so the same rules as always should be applied with regards to protection. The risk of pregnancy depends on if your period falls in your fertile window or not. That is, however, highly unlikely unless you have very short cycles.

How much should you bleed?

Researching the quantity of menstrual blood seems to be a whole science on its own. Up to now, most researchers have chosen the strategy of using tampons and pads to estimate the blood loss, rather than systems such as cups. Depending on their type and how much they are filled, they are counted as a certain number of millilitres, which is most likely not very precise.

The main factor determining the amount of blood loss has to do with the quality of the uterine lining. If we bleed too little, the reason may be that not enough progesterone is produced to properly protect the uterine lining from inflammation. This means that the menstruation will happen earlier meaning that the lining was less built up. Since oestrogen also plays a key role in building up the lining, its amount is equally important. Besides, we tend to bleed more during the first fertile years just after the first menses as well as during the very last fertile years before menopause. This is because ovulation is irregular in these phases and the intervals between the bleedings are longer. When the lining had time to build up over a longer time, the bleeding will be more important when it comes.

What does all this mean in terms of actual quantities? In a 2012 study where 201 women were followed over two cycles,[90] scientists found that the majority of the women bled between 3 to 7 days, with heavier bleeding during the first 3 days (and the second one in particular). The amount of blood loss went from 15 to 114 ml, but the average was 45 ml (about an eggcup).

The maximal limit for normal uterine bleeding is 80 ml – half a tennis ball – per menstrual cycle.[91] There might be many reasons behind heavy bleeding, the most common ones being hormonal imbalance caused by polycystic ovary syndrome (PCOS), obesity, insulin resistance or thyroid issues. Other reasons include uterine fibroids (benign tumours inside the uterus), endometriosis (when the endometrium is growing outside of the uterus), and adenomyosis (when the endometrium grows inside the muscle wall of the uterus).

It is estimated that one in ten women suffer from heavy bleeding, a predicament that can be quite debilitating in their daily lives. It also has a significant effect on society as a whole. In the UK, it is estimated that the direct cost of this problem in terms of treatment for the healthcare system is £100 million, with the indirect cost in terms of not showing up to work being as large as £1.8 billion.[92]

Chance-box

The length of the menstrual cycle and the length of bleeding is also linked to the chances of getting pregnant. In a study of 490 women, the authors found out that 30–31-day cycles and 5 days of bleeding were the typical cycle pattern of women who got pregnant the fastest.[93] They also had a lower risk of spontaneous abortion. This is, however, a topic that researchers do not agree on, some studies have confirmed this,[94] but another found that cycles of 28 days were the most likely to lead to pregnancy.[95] When reading such information, it is important to keep in mind that these are observations on groups of women and for you as an individual it will not mean much.

What should it look like?

In the beginning of the bleeding, it takes a longer time for the blood to exit the uterus. This makes the blood darker. Once the outer layer of the uterine wall is shed, there is an open wound that bleeds directly out of the uterus: the blood

Quantity of menstrual blood
based on...

Feminine pads

TYPE

DAY ☼	1mL	2mL	3mL	4mL
NIGHT ☾	1mL	3mL	6mL	10mL

Tampons

TYPE

LIGHT	0,25mL	0,5mL	1mL	3mL
MEDIUM	0,5mL	1mL	1,5mL	4mL
HEAVY	1mL	1,5mL	3mL	8mL
SUPER	1mL	2mL	4mL	12mL

becomes a clearer red. Towards the end of the cycle, the flow slows down again leaving more time for the blood to exit and hence react with the oxygen to change back to a darker colour.

When should we bleed?

All vaginal bleeding is not menstrual bleeding. Menstrual bleeding is always a consequence of ovulation. All other bleedings are not menstrual bleedings. When using hormonal birth control, the bleeding is not a real menstrual bleeding but a so-called withdrawal bleeding.

When the bleeding is outside of the normal menstruation period, it is called spotting. It can sometimes be a bit difficult to define spotting. A small bleeding within 2 days of your normal period is not considered spotting but a part of the normal bleeding. Spotting outside of this window, in the middle of the cycle, should normally not occur, but on a few occasions, it can happen without being a sign that anything is wrong. This is during the first months of taking a hormonal contraceptive, or when you forget to take the pill. It can also happen during ovulation and early pregnancy as well as in perimenopause. It is called a breakthrough or withdrawal bleeding and is linked to imbalances between or abrupt changes in progesterone and oestrogen.[96] Spotting seems, however, to be rather rare. In a group of 201 women followed over two cycles, only few women experienced spotting, 4.8%, and in only 2.8% of cycles.[97]

Period pain

Dysmenorrhea, or period pain, has previously been considered normal and partly a psychological issue, but is now, finally, being taken seriously. It has been shown that women with strong period pain have a much larger number of prostaglandins that contract the uterus than other women.[98] Prostaglandins are a necessary part of getting rid of the menstrual blood, and since it causes the uterus to contract stronger, it's also the most likely reason for period pain.[99] If the amount of prostaglandins is high, the pain can also radiate out from the lower abdomen and cause lower back pain. Dysmenorrhea is divided in two sub-groups, primary and secondary dysmenorrhea. Primary dysmenorrhea is pain without a related problem and secondary is linked to an underlying disease.

Strong period pains should never be considered normal, and it is important to understand the underlying cause. Today's healthcare has become much better in taking this problem seriously and finding its causes, such as endometriosis, more quickly. Only two decades ago, it took on average 12 years between the first symptoms and the diagnosis in the US and 8 years in the UK.[100] Luckily, this time has been reduced significantly over the last few years.[101]

As mentioned earlier, prostaglandins play an important role in the shedding of the uterine lining, which is why NSAIDs, such as ibuprofen or naproxen, that work actively against prostaglandins, can both relieve the pain and reduce the menstrual flow.[102]

Is the blood loss bad?

A few years after the birth of my second child, I got more and more tired. Even though the fatigue was so bad that I had to hold on to the kitchen sink to keep upright in the mornings, I thought it was normal. After all, I was a working mother with two small children at home. I did not seek help, but one day I needed to go to the doctor for another issue, and he asked me how I felt otherwise. "Maybe a little tired," I told him. Since I was used to doctors giving me a pad on the back and telling me that it would pass, imagine my surprise when he asked for more details. It all ended up with a blood test, where it was settled that I had very low ferritin levels. Ferritin is a protein which stores iron in the cells and, some of it being released in the serum, measuring the amount of ferritin in blood serum is a reliable way of detecting iron deficiency anaemia.[103] Iron deficiency anaemia is a decrease in red blood cells or the haemoglobin due to lack of iron, which is an important component in many cells, mostly in the blood cells, where it helps transport oxygen. There is, however, a stage before you reach anaemia, which is when you are more or less depleted of your iron storage but still have sufficient amounts of haemoglobin. This is called latent iron deficiency and is not as bad as anaemia but can lead to a very severe fatigue.[104]

My doctor gave me an intravenous infusion of iron, and it completely changed my life. A few days after, I would jump out of bed in the mornings, and that para- lyzing feeling of not being able to move my body properly was completely gone.

So, what caused my iron deficiency? My doctor told me that it was because of menstrual bleeding and that was common in women. However, studies have shown that a normal menstrual bleeding should not cause iron deficiency.[105] From an evolutionary standpoint, it would not make sense for our bodies not to be able to cope with our natural functions. More likely is that my two preg- nancies had pushed my iron levels to the minimum. During pregnancy you use more iron to create more blood both for you and the baby, and iron deficiency is very common and regularly checked during pregnancy. About 6 months after the delivery of my second child, I got a copper intrauterine device (IUD) implan- ted, which is known to create heavier bleeding. Women at the lower end of iron storage sometimes do get ferritin and haemoglobin deficiency because of the copper IUD.[106] A normal menstruation bleeding should not be a reason for anaemia per se, but the fact remains that a very large portion of menstruating women suffer from iron deficiency, leading or not to anaemia. Iron deficiency is the most common nutritional disorder, with menstruating women being the most affected. In the UK, it affects around 20% of all menstruating women, for 33% in Switzerland, depending on the definition.[107]

So, no matter what the reasons behind the iron deficiency might be, if you suffer from symptoms like severe fatigue, weakness, brain fog, muscle, and joint pain, it can be a good idea to have it checked.

Toxic shock syndrome

When I had just gotten my first period, my mother very actively tried to talk me out of using tampons, because she was so worried that I would get something called toxic shock syndrome (TSS). The toxic shock is a very sudden and strong reaction of the body to a certain bacterium which can be found in many women's bodies. The blood pressure drops sharply and cuts of the oxygen supply to vital organs which is why it can be potentially lethal. Nevertheless, I did start using them, but I was very careful to always change them every 4 hours, because I was convinced that I would immediately get TSS. My mother's worry was probably well founded back then. TSS was named in the late 1970s when the incidence increased. In Canada, 53 women died of it between 1976 and 1981, and 1200 women in the US during the same time period. All those deaths were mainly linked to using a specific type of tampon. As soon as that tampon was withdrawn from the market, the number of deaths rapidly declined.[108] The reason for the TSS was a bacterium that in some rare cases propagates when using a tampon for too long. Nowadays, incidences of TSS are very rare, about 1 case per 100,000 women per year. However, the recommendation is to use the least reabsorbing tampons to make sure that you change them every 4 to 8 hours.

Menstrual blood as a diagnostic tool

As you have seen, simply analysing the amount and appearance of the menstrual bleeding can give you a good indication of your health. Interesting research is ongoing on how proteomics could be used to analyse the liquid in order to understand underlying issues. Proteomics is the study of the proteome, i.e., the entire set of proteins produced by an organism. If the genome is the construction plan of the body, the proteome shows how that plan is actually implemented. Some researchers have been working on establishing the proteome of menstrual fluid which could potentially give clues on the causes behind infertility.[109]

Attempts have been made to develop a test for detection of endometriosis from the menstrual blood. There is no solution so far, but the use of proteomics is promising.[110,111]

CYCLIC IMPACT OF HORMONES

The knowledge that we are in many ways influenced by the cyclic changes of our hormones is not new. The problem in the past was rather that their influence was terribly exaggerated, suggesting that women were the victims of the hormonal fluctuations to such an extent that they were unable to control themselves. In the nineteenth century and early twentieth century, some women were even acquitted for murder by claiming menstruation issues. In 1845, a domestic aid got away with murdering one of the children of her employer by claiming insanity due to "obstructed menstruation."[112] This was a term used to define women who, for some reason, did not menstruate; most medical practitioners of this period were convinced that a woman who failed to menstruate would suffer from an even more extensive list of symptoms than those characterizing menstruating women.[113] In 1851, another woman was freed after having killed her baby niece, the excuse being this time the insanity raised from "disordered menstruation." Smaller crimes, such as shoplifting and theft, also could be pardoned due to menstrual issues.[114]

In 1984, some scientists were still claiming that the menstrual phase was linked to crime, bad academic performance, alcoholism and other 'antisocial behaviour'.[115] The first British professional boxer, Jane Couch, had to sue the British Boxing Board of Control for sexual discrimination to get her license to box: it was not the nineteenth century, but 1998. The professional association had claimed that premenstrual syndrome made women "too unstable to box."[116] It would have been nice to be able to wave off some of your own insanity to menstrual issues, but it is probably a good thing that this view has changed. Newer, more carefully designed studies have been able to debunk the myth of women being entirely controlled by their hormones. Considering that prejudices being around for such a long time and with such deep and widespread roots, one is tempted to assert that there must be some truth to it, but keep in mind that they came up in very misogynistic contexts, even present in science, with its powerful expectation bias.

Even though the old ideas were largely exaggerated, we are influenced by our hormones in many ways. The hormones, produced in the glands and in the various cells, travel through our bodies within the bloodstream and trigger reactions wherever they encounter cells that are ready to receive them. The cells that can be influenced by the hormones have specific receptors for this purpose, and they can be found everywhere in the body. This applies to sex hormones, oestrogen and progesterone produced by the ovaries during the menstrual cycle as well. Apart from the obvious places linked to reproduction, receptors for those hormones can be found in most tissues of the body, such as the eyes, the skin, the bones, different parts of the brain and the gastrointestinal tract.[117] No wonder that the cyclic changes of the hormones have many and sometimes unexpected effects.

I am not in favour of letting the hormones limit me in any way, but I admit that when I found out that my TEDx talk would take place during my fertile phase – a time when I feel more confident and radiant – I was very happy. A friend of mine told me that she tries, if possible, to book her media interviews in her follicular phase for the same reasons. It does not mean that we cannot perform during other phases, but we do it with more ease at some moments.

Remember, when reading this chapter, that this is the state of current research and many things are yet to be discovered. If you experience something during your cycle on which there is no current research, it does not mean that your experience is an illusion, it only means that the research on it has not been performed yet. Tracking your own data can be useful in figuring out how it might be linked to your hormones or if it is rather a random symptom. If it is linked to your hormones, the effect will then recur with a cyclic regularity, meaning that it is no longer unpredictable.

Brain – improved wiring

During my PhD, I was developing methods to analyse a new type of images that could map the brain connectivity using magnetic resonance imaging. As we were always lacking images to work with, my office mate and I had to spend many hours in the scanners, producing images of our brains. Inevitably, we spent a lot of time analysing our own brains and one of the things we found out was that my two brain hemispheres were better connected than his. Since he was both more serious than me and a higher performer, I relished in interpreting this fact to my advantage. Better connectivity simply had to mean that I had a much broader view on life and could handle both the intellectual and emotional side of things even though I sometimes struggled to stay focused on only one topic. Since my colleague was not only a very smart guy but also very polite and diplomatic, he agreed to all my preposterous claims. There are many speculations on what exactly such differences in brain structure mean, but it is interesting to notice that oestrogen and progesterone have been found to improve the communication between the left and right brain hemispheres,[118] so our observations were not entirely random. We were indeed representative of the statistical sex averages with regards to brain structures. The connectivity between the two hemispheres might even change during the menstrual cycle with the fluctuating hormones.[119] Some researchers have speculated that this alteration in connectivity over the month could be beneficial for the ability to look at a problem from different points of view.[120]

It is always tricky to deal with gender stereotypes. Especially when it comes to the brain, which tends to be a very sensitive subject. It is important to remember that we talk about differences on average, meaning the average woman compared with the average man. Knowing about differences in average will not provide scientific ammunition to judge one's capacity based on sex, neither in a positive nor negative way. It is, however, interesting to study differences as they can help us understand how the hormones are influencing the body.

Hormones also influence the wiring of the brain. The brain is full of receptors for both progesterone and oestrogen. The receptors stimulate how the brain continuously changes and adapts to its environment by producing more myelin. Myelin is the insulating fatty layer surrounding the nerves in the brain which facilitates the signals, transmitted via the nerves, to travel faster and more efficiently. Myelin can be seen as the barriers surrounding a highway, preventing cars going off the road and preventing animals crossing it; these barriers allow the cars to go much faster and in the right direction. The myelin gives its colour to the white matter, which is the long-distance wiring in the brain, and helps different parts of the grey matter to communicate. It has been shown that progesterone plays an important role for the brain to recover after brain injuries through the stimulation of myelin growth.[121,122] Both progesterone and oestrogen slows down the cognitive decline that is linked to aging and protects against some neurodegenerative diseases, such as Alzheimer's or Parkinson's disease.[123, 124] There are also indications that it can attenuate the extent of acute injury to the brain that is caused by stroke and brain trauma.[125] This effect protects women while they are still in their fertile years, but once they have passed menopause, they have a much higher risk of Alzheimer's than men. The mechanism is not completely understood yet, but it is likely a combination between genetics, hormones, and social situation.

If the white matter of the brain is where the information is mainly transferred, the grey matter is where the information is being processed. Agatha Christie's fictional detective Hercule Poirot always referred to "those little grey cells" when it came to using his intelligence to solve a crime and amount of grey matter in some regions is indeed linked to intelligence. The grey matter is mainly composed of neurons. It is also influenced by hormones and it has been shown that grey matter changes structure with the menstrual cycle.[126] These variations differ between naturally cycling women and women using oral contraceptives.[127, 128] Despite those differences, there is no indication that our ability to solve complex problems and think properly varies with the cycle.[129]

Hormones also have a strong influence directly on the structures in the brain. We already know how it influences the hypothalamus as a part of the reproduction, but the hypothalamus is also the control centre of many important bodily functions such as hunger, thirst, temperature, water level, salt level and much more. The amygdala and hippocampus are two other brain structures which have many receptors for progesterone and oestrogen.[130] The hippocampus is important for storing memories which might be crucial for social abilities. When you remember how you felt in a certain situation, it is much easier to relate to others and develop empathy. Some studies show that the hippocampus get more oestrogen receptors in certain phases of the menstrual cycles.[131]

The amygdala is instrumental in the process of emotions, especially fear and the decision to fight or flee. This helps us to have a better understanding of how other people feel. When women are in their fertile phase, the amygdala is under the stimuli of oestrogen, which some researchers have linked to being more careful. Anthropologists have speculated that being more fearful in this phase might be a way to avoid situations that could lead to rape during the fertile period.[132, 133] This finding is, however, in contradiction with the fact that testosterone peaks during this phase, and that hormone is normally a driver of risky behaviour in women.[134, 135] Here is one of the many examples of how difficult it is to get clear answers when it comes to the human body.

If scientists are right in thinking that the oestrogen is pushing for these changes in the amygdala and hippocampus, it could explain the stereotype concept of women being more social and empathetic on average.[136] Again, pay attention to the on average. Many factors determine our behaviour, social environment, and heritage, i.e. that the hormonal influence is just one small part of the puzzle.

Another unexpected effect of hormones during the menstrual cycle is that it influences your capacity of changing habits, such as quitting smoking which is easier in some phases. In animal models researchers have shown that the reward system for nicotine and other drugs is higher in phases with elevated oestrogen levels and lower with raised progesterone level. Similar findings have been found in smokers so the conclusion is that quitting should be easier in the luteal phase of the menstrual cycle.[137]

Mood – the cyclic ups and downs

You might feel that your mood is not the same over the whole cycle. On the days just before the next menses, the mood may feel down, but overall, there are more positive than negative effects of the hormones over the cycle.

Mainly due to my professional activity, I have paid great attention to how my mood fluctuates throughout the month and my cyclic patterns are quite clear. Just after menstruation stops, I feel great, sociable and lively. And hornier. This phase lasts up until the ovulation. A few days after ovulation, my emotional state transforms into something calmer, where I feel like staying home and being cosy with my family. Then, about two days before the next menstruation, I am possessed by a great wave of insecurity: I think that the world hates me, and a simple dry look is enough to convince me that the whole town considers me worthless.

These kinds of feelings over the cycle are not the same for all women but the general traits are quite common, and they can be traced back to hormonal changes. On average, oestrogen stimulates the feeling of happiness and progesterone calms you down. When suddenly all these effects disappear as the level of hormones is diminished towards the end of the cycle, women can react more or less strongly.

When oestrogen acts like natural Prozac

The way oestrogen acts and makes you happy is explained by an increase of the serotonin level in the brain. Serotonin is a neurotransmitter which has many effects, but it is better known for improving our feelings of happiness and well-being. Serotonin can be seen as factory workers producing more happiness, but they do not want to work overtime. The principal effect of many antidepressants – selective serotonin reuptake inhibitors (SSRIs) such as Prozac for instance – is to lock the doors of the 'happiness factory', preventing the workers from getting out and thereby forcing them to do overtime.[138] This class of drugs block the neurons from reabsorbing the available serotonin, leaving more serotonin for us to make us feel happy. Oestrogen acts to a certain extent as the brain's natural Prozac by stimulating the production of serotonin and making sure it does not metabolize too quickly and hence stays longer in the brain.

The oestrogen suddenly dropping towards the end of the cycle impacts the amount of serotonin and hence our mood, and even otherwise healthy women report lower moods and negative emotions.[139] The trend is such that a correlation between suicide attempts and low oestrogen levels in cycle has been revealed.[140]

Since oestrogen plays such an important part in the serotonin production, its sudden variations of level have a huge impact not only during the monthly cycle, but on two other occasions as well: the days after giving birth and the menopause, periods witnessing even bigger drops in oestrogen. Just after giving birth, when the placenta is expulsed, the oestrogen quickly plummets from about a hundred times the normal concentration to very low within a couple of days.[141, 142] This sudden drop in oestrogen leads to the increase of an enzyme (enzymes are molecules that speed up a reaction) that metabolizes serotonin. It means that serotonin level decreases leading to potent mood changes just after birth.[143] For most women this is only a temporary state. Women with postpartum depression have a high concentration of this enzyme:[144] hence, their serotonin levels stay low over a longer period of time which explains their emotional state.

In menopause, which also causes a strong drop in oestrogen levels, some mood disorders become more common.[145] Especially early menopause can be a trigger of depression,[146] but here is a chicken–egg situation because some women with previous depression may have a higher incidence of early menopause.[147] Evidently, the dropping hormones is unlikely to be the only cause for depression. Those periods with large hormonal changes often occur simultaneously with other changes in the life situation. During those key moments in life, physical, psychological, and social layers of stress pile up making it a very challenging time.

However, it is also common that women become in a better mood during menopause. Indeed, the lacking oestrogen might be compensated by the sudden disappearance of the strain associated with the cyclic variations.

It is important to note that oestrogen is not the only thing needed to produce serotonin, another important component being vitamin D. Vitamin D is produced in the skin when it is exposed to sunlight; having too little of it may cause depression in general and especially during the winter season if you live far from the equator.[148] If you live in a place where getting out in the sun is not an option – like Sweden where I come from – supplements are available. Choosing the right food can also boost the production of serotonin since it is synthesized from an amino acid called tryptophan, which is found in foods like meat and fish, dairy and tofu, nuts and seeds.[149]

Up to now, all subjects dealt with have been related to biochemistry, but the chemistry is of course also influenced by the social and personal context. We have talked about sun and food, but stress linked to your life situation will raise the level of another hormone, cortisol. High cortisol levels will stimulate the very enzyme that breaks down serotonin, so it will further reduce the serotonin levels.[150] Hence, stress and life situations can make premenstrual mood swings much worse and trigger postpartum and menopause depression.

The natural calming effect of progesterone

The effects of progesterone can be a bit confusing. Progesterone has a general effect on the amygdala, the centre of the brain and regulator of emotions such as anger, fear, and anxiety. At some concentrations, progesterone has a calming effect on our mood. It can reduce feelings of anxiety and release tension and stress and help against depression. It can also have a slight sedative effect and make you feel more tired. During phases when you do not have much progesterone, it will be produced by the adrenal gland as a response to stress. The hypothalamus comes once again to the rescue, triggering this response to help us cope with the increased stress.

It has been shown that progesterone may increase a desire to be close to others. This effect is beneficial during times of stress, as social support makes us feel better.[151] Too little progesterone can make some women aggressive and tired in the premenstrual phase.[152] On the other hand, having too much progesterone can also have a depressive effect as progesterone seems to set off the amygdala, the part of the brain that triggers fear and anxiety.[153]

There are many ways to calm down after a stressful day and – whether you do it with a glass of wine, yoga[154] or by taking a Xanax – all those ways will act on a special type of receptors in the brain (GABA-A receptors) that will activate a reaction that calms you down. Progesterone (or actually its metabolite, allo-pregnanolone) also bind to these receptors, and it is the attachment that will set off the reaction.

If you are used to taking a glass of wine or a Xanax to relax and then suddenly stop, this can cause anxiety: a similar thing happens at the very end of your menstrual cycle.

The 'more-info-than-you-asked-for'-box

What makes progesterone a bit special is that in some women it does not only bind to the receptors, it also changes the sensitivity of the receptor, a fact that can create a negative reaction to progesterone. This can alter how you react to other stimuli, for instance to stress. This effect may lie behind the most severe cases of premenstrual syndrome (PMS), known as premenstrual dysphoric disorder (PMDD). Women with PMDD and also with burnout symptoms seem to have an altered reaction in their GABA A receptors which makes them react negatively to progesterone.'[155]

Premenstrual syndrome and premenstrual dysphoric disorder

My husband asked me recently if he could download my menstrual tracking app so that he could be more up to date with my latest mental status. Likewise, the questions men generally ask me pertain mood swings and premenstrual syndrome (PMS). What is it, why is it happening, how can you tell if it is coming, is it real? Whether it is a monthly reality for women or not, it has become a sincere concern among men. Some claim PMS is a social construction, because it seems that women with more difficult social and domestic situations suffer more than others, and that treating PMS as a physical symptom is dismissive of their experience. The hormones – with its variation throughout the menstrual cycle – are definitely at work, but the life-situation is highly influential too creating chemical reactions in the brain: both components can worsen the symptoms or alleviate them. To give another personal example, I experienced a very strong PMS during the first month of the lockdown during the coronavirus pandemic in March 2020. For us, as for everyone else, it was a very stressful and uncertain time with a lot of changes to adapt to. Once we had taken our marks and settled down – with a life even easier in some respects – the following PMS disappeared almost entirely. The fact that a reaction is chemical does not make your experience, and whatever external event that triggered it, any less real.

Feeling understood during the period of PMS also seems to have a positive effect on how we experience it. A friend, who lives in a relationship with another woman, swears that the empathy you receive when being entirely understood can dampen the symptoms tremendously, an assertion supported by studies on women in lesbian relationships.[156] It makes sense as comprehension and empathy lower stress and cortisol levels, which in turn prevents a fast serotonin breakdown.

About 3–8% of women in their fertile age are struck by mood swings of such a severe level that it heavily impacts on their quality of life and can even jeopardize/put in peril their relationship. In these cases, we are not talking about PMS

Pathways for PMS

(premenstrual syndrome)

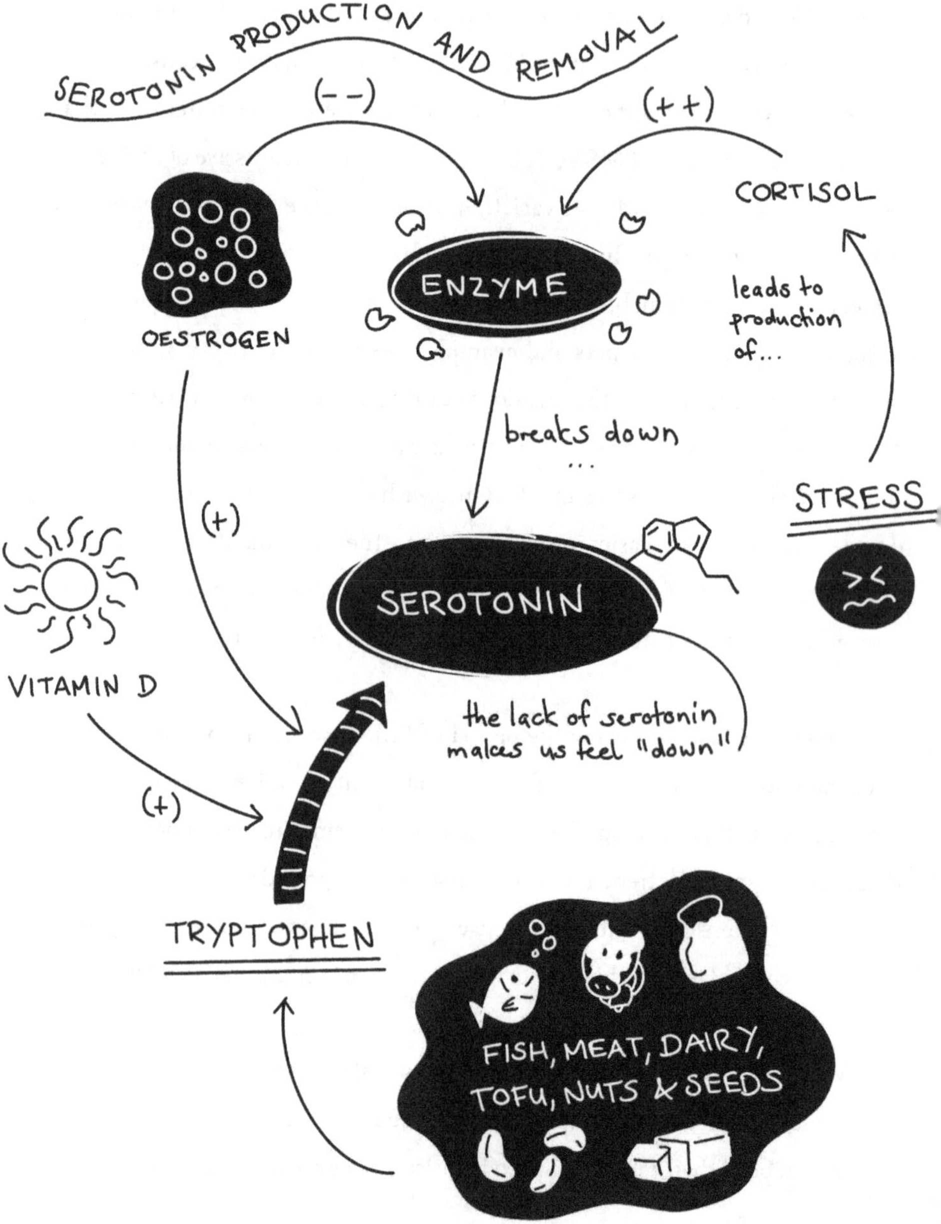

anymore, but premenstrual dysphoric disorder (PMDD). Having PMDD means that you become terribly irritated, depressed, aggressive and emotionally labile.[157] To be diagnosed with PMDD, you need to have at least five of the following symptoms: depression, irritability, anxiety/tension, affect lability, decreased interest, difficulty in concentrating, fatigue, feeling out of control, insomnia, change in appetite, breast tenderness or breast swelling, with at least one of these symptoms being a mood symptom. The symptoms must also be severe enough to interfere with usual activities and they should only be present in the second phase of the menstrual cycle, the luteal phase.[158] If they are present over the entire menstrual cycle, they are linked to something else. The difference between PMS and PMDD is that the latter has much stronger mental symptoms and the patients suffer for a longer time. PMS occurs only the days preceding menstruation, during the hormonal withdrawal, but women with PMDD suffer during the entire luteal phase.

PMS and PMDD are both linked to the changing hormones but in different ways. The milder form of mood swings, as in PMS, are not so much linked to the hormones themselves as to the lack of hormones, and especially to the lack of serotonin when oestrogen is withdrawing.

In PMDD, progesterone plays a stronger role. Researches are trying to figure out exactly what happens, but the most popular hypothesis is linked to how the progesterone – or more precisely allopregnanolone – changes the receptors we mentioned before, the GABA-A receptors.[159] When allopregnanolone binds to the receptors, it changes their function and, in the case of women with PMDD, the effect is particularly strong and changes the sensitivity of these receptors during the luteal phase.[160] When the sensitivity of the receptors is modified, it makes it more difficult to calm down, which leads to anxiety and irritability. The effect increases as the amount of progesterone increases, but the strongest issues occur when the progesterone withdraws. This is why PMDD symptoms are the strongest during the 2–3 days preceding menses and relieved during menses.[161]

> **The PPP-box (Preventing PMDD with Progesterone)**
>
> It has been shown that PMDD only occurs in ovulatory cycles,[162] which means when progesterone is triggered. Hence, suppressing ovulation would help against PMS/PMDD but not necessarily when it is done by using oral contraception (OC). As a general rule, oral contraception is not great for mood. About half of all women that stops using OC name negative mood changes as the main reason,[163] and women who suffer from PMS are also more likely to have negative effects from hormonal contraceptives. As hormonal contraceptives contain different synthetic versions of progesterone, they all have different effects on the brain.[164] Providing extra progesterone can help against both PMS and PMDD.[165]

Some scientists assert that postpartum depression is linked to the same phenomenon, i.e., a change in the receptor function. In the beginning of a pregnancy the receptors are working normally which is why, during the first months, the increasing progesterone can cause a lot of symptoms such as sleepiness etc. After that, they get down-regulated, which is a positive thing while the levels are high but might lead to depression when the retraction of hormones is too rapid which happens at the end of the pregnancy.

So, is there something else you can do to soothe hormonal symptoms and mood swings? We mentioned vitamin D supplements and the reduction of stress, and there are also indications that calcium and magnesium can help against many of the symptoms.[166] In addition to helping with mood swings, it can also reduce period pain and menstrual migraines.[167] Magnesium in combination with vitamin B6 seem to have the best results.[168] Previously, findings indicated that coffee might worsen the PMS symptoms but to my personal great pleasure one study contradicts that.[169] It has also been shown that women with a BMI over 25 suffer more from PMS.[170] However, it is not clear if it is an effect of the weight per se or if it is due to the increased stress one might feel due to the social stigma that often comes with having overweight.

An evolutionary purpose of PMS?

Menstrual research is not only about investigating what happens in the body from a hormonal perspective, but also about trying to understand the evolutionary sense of what is happening. Why was the evolution of PMS allowed to happen in the first place? One interesting theory is that the purpose of PMS is to drive away partners who were unable to make you pregnant. Since you only get PMS at the end of your cycle if you are not pregnant, your partner has obviously failed doing his job. From an evolutionary perspective, he has therefore failed in his main purpose and it would make evolutionary sense to try to get rid of him. The reasoning behind this argument is based on a finding that PMS seems to be primarily directed towards partners,[171] but not everyone agrees with this idea.

The evolutionary purpose of PMS is altogether questionable. Before modern times, we would only menstruate around 100 times in a lifetime, since we would be pregnant most of the time, which is why PMS would not matter so much. Hence, PMS could simply be an evolutionary accident, with a much greater impact on life nowadays, while women menstruate about 400 times during reproductive life.

All the other not so pleasant PMS symptoms

PMS is, however, not only about mood swings: PMS includes a much wider set of physical and psychological signs with up to 150 different symptoms. It covers physical symptoms such as breast tenderness, bloating, acne, abdominal pain, fatigue, as well as psychological symptoms such as being overly emotional, anxiety and depression. About 80% of all women experience some kind of PMS symptoms although most of them are mild, whereas 30–40% have symptoms that require some form of treatment.[172] Symptoms start getting worse about one week before menstruation and peaks 2 days before menses. All of them have a different cause.

Some of them are triggered by prostaglandins which are the local hormones released when the uterus is shedding its lining. We already talked about how they can create menstrual cramps, but they also play a role in worsening other symptoms such as asthma. For 20–40% of women with asthma, the symptoms get worse during menstruation.[173] They are also known to trigger headaches. Since migraines are linked to dropping levels of oestrogen, they tend to increase during the premenstrual phase.[174]

If you suffer from epilepsy, you might benefit from some relief during your luteal phase, as the metabolized form of progesterone, allopregnanolone, has an anti-epileptic effect. However, just as for PMDD, some women develop tolerance to allopregnanolone, and at the end of the cycle and during menses they get a withdrawal effect when the hormone is reduced, which leads to more seizures.[175]

Let's talk about sex

In my late twenties and early thirties, I was completely obsessed with the idea of having children. I could think of very little else, and the thought of babies completely dominated my life. If someone had asked me then if this urge was biological, I would have said yes without hesitation, because it felt completely beyond my control. There is, however, no scientific evidence that the desire to have children is biological. Although many women, and men for that matter, feel a desire to have children, this may partly come from culture, social pressure and urge to be considered successful.[176]

The sex drive is crucial for our existence and, until very recently, sex performed during the fertile phase would lead to pregnancy whether you desired it or not. Hence, there is no need for a biological wish to procreate. But, for the survival of mankind, there is a necessity for sexual desire and pleasure.

Lust – a complex cocktail

Hormones do play an important role in sex drive, but the causality is not so obvious as to enabling scientists to assert that women are aroused during the fertile phase. Sexuality is a much more complex matter, involving a large battery of biological, psychological, or social components. One particular problem in studying human behaviour is that we, humans, are too aware of the consequences of our actions to fully cede to our natural instincts. Today we are influenced by a great number of external factors: the society we live in, the perceived expectations, the stress in our daily lives, our current feelings towards our partner (if we have one) and many more. This makes it difficult to study a 'feeling of being aroused' in general or during the fertile phase in particular.

However, there seem to be such a tendency: studies – having tried to compensate for the social environment and other variables – found out that it was more common for women to take initiative to sex during the days before ovulation.[177] There is also evidence that oestrogen increases the sex drive, since two days after the oestrogen level rises, the sex drive increases.[178] A sudden peak of testosterone in that phase will further increase the sex drive. Women do seem to seek more contact with men in their fertile phase[179] and are more likely to give away their phone number in this phase.[180] Even though there are plenty of studies on sexual behaviour and Tinder – a dating application focusing on short term relationships – none of them focused on the menstrual cycle.

Progesterone has a negative effect on the libido[181] and 2 days after the progesterone level rises, the sex drive plummets, and higher progesterone levels lead to less sex. This can also be one of the negative side effects of the birth control pill that often has a high dose of artificial progesterone. On the other hand, the commitment to your partner seems to rise, when the progesterone level is high.[182] The negative effect of progesterone on libido may be the reason why many women suddenly find back their lust just before and during menstruation. It might also be the reason why some women experience more libido in menopause; they have lost the positive effect of the oestrogen, but the negative effect of progesterone is also gone.

As we have seen, sexual motivation comes from a complex cocktail of hormones. Testosterone is the main driver for men, but also is a strong influencer in female lust. The testosterone is increasing around mid-cycle, can lead to an even stronger sexual drive just before ovulation. In menopause, the sexual lust can go in any direction. Some women completely lose their libido due to the oestrogen levels dropping, whereas other women actually can feel an increase in libido since the amount of freely circulating testosterone increases just after the menopause.

Oxytocin is a hormone that is released when we are close to other people and it is an important hormone for how we bond to our babies. For women, it is also important in sexual lust. It is not only released when breastfeeding but also in high doses during orgasm.[183] Also prolactin, another hormone that stimulates breastfeeding, increases after an orgasm. The fact that breastfeeding and orgasms release the same hormones does not mean that their effects make you feel the same way. Believe me, they do not.

Orgasms – for the survival of mankind

Considering how important sexual motivation is for the survival of mankind, it is only natural that we are supposed to enjoy it and want to do it again. Unsurprisingly, orgasms are important for our motivation, so it is not only important that it feels good, but we also need to remember what an amazing time we had. Orgasms release a whole cascade of different chemicals which all have a different purpose. Dopamine is one of the neurotransmitters released during orgasm and is known as the 'feel-good chemical'. But it also plays a big role in learning, so it helps us remember that we like having sex, which is useful for the long-term lust.[184] Another hormone released during an orgasm is serotonin, which makes us feel happy; it is the combination of oxytocin and serotonin that make us feel drowsy after sex.

Pleasure-box

Orgasms are more likely during the fertile window as the clitoris grows significantly during this phase.[185]

The release of oxytocin, serotonin, dopamine and also prolactin during orgasm is the same for both men and women. However, in the female brain, oxytocin keeps being produced also for a while after sex, which might explain the stereotypical image of men wanting to sleep after sex with women wanting to cuddle.

Not everyone is convinced that sexual motivation is a sufficiently strong evolutionary reason to why women have kept their capacity of orgasm. I say 'kept' because the clitoris – which is the organ that generates the orgasms – stems from the same structure as the penis, reason why, when you look at the entire inner structure of the clitoris, you can see that men's reproductive system resembles. It is therefore the same mechanisms that are important for orgasms in both genders. A vaginal orgasm is rather a biproduct of accidental clitoris stimulation.

As ejaculation is necessary for procreation, it is rather obvious why men have orgasms but even in women the capacity of having an orgasm must have a more important purpose, otherwise that whole structure would have disappeared through evolution since an orgasm consumes a lot of energy. There is an interesting 'upsuck' theory, which is the theory that more sperm is drawn into the uterus if we have orgasm.[186] This theory seems to work well in pigs, as Mary Roach wonderfully shares in her TED-talk,[187] but in humans an orgasm is in no way necessary to procreate. There is also no link between the number of children and the number of orgasms a woman experienced. Instead, it is believed that orgasms have a cultural importance and can act as a social glue and that mutual pleasure can create stronger groups. Something that was important for our ancestors and can still be observed in the bonobos, a primate species known to solve all their conflicts with sex.[188]

Among the many theories explaining female orgasm, one points to ovulation. In some mammals, the orgasm serves as a trigger for ovulation, and it is possible that this is where the orgasm evolved from. However, at some point in our evolution, we started having spontaneous ovulation which could be a consequence of the clitoris moving further away from the 'copulation canal', or the vagina, as technically – and quite lyrically – referred to by scientists.[189]

Proximity – who we want to cuddle with and when

Beyond sex, hormones have other influences on our sexual behaviour, and some studies claim that we feel differently about our partners depending on where we are in our menstrual cycle. Women with a partner they found sexy felt closer to him or her during their fertile phase, whereas women with a partner they did not find as attractive felt more intimate during the luteal phase.[190] Hence, the menstrual cycle can impact the choice of partner: it seems that we are searching for more attractive partners when the oestrogen level is high. Being on the pill might also influence your choice of partner, but since each pill has a different composition it makes the matter even more complex to assess.[191]

Metabolism – All about the gut-feeling

Appetite – from balanced to binge eating

All my attempts to eat less somehow fails after a week or two. It seems that my body is having its own way. And in some ways, it is actually true. The appetite does change significantly throughout the phases of the menstrual cycle. Scientists have shown that the fluctuating levels of hormones impact appetite both directly and indirectly according to the increased energy needs connected to the bodily changes induced by the hormones. We mentioned earlier that in the luteal phase, when we have built up the uterine lining, our energy needs an increase by about 7%. This might explain why we tend to eat more during the luteal phase than during the follicular phase.[192] Some studies report that women eat as much as 90 to 500 kcal more per day in their luteal phase. Oestrogen increases our energy consumption, but not to the same extent as testosterone does, which has three times higher effect than oestrogen.[193]

The quantity of food we eat is not only determined by how much energy we consume. Appetite is what decides how much we eat, and oestrogen plays a strong role there. By balancing the appetite, it inhibits our food intake, so we eat less.[194] Taking additional oestrogen in menopause sometimes reduce the weight gain that often follows the transition to menopause. Oestrogen also seems to limit emotional eating, and this is why we tend to do more binge eating during the luteal phase.[195] Progesterone alone does not increase the appetite, but it seems different when it interacts with oestrogen as in the luteal phase.

Diabetic-box

If you are diabetic, it is important to understand the hormonal influence on the glucose-level reaction to the administered insulin.[196] Indeed, oestrogen and progesterone have different effects on the cells' sensitivity to insulin. The insulin sensitivity decreases during the luteal phase due to the progesterone rise and increases during the follicular phase, which is most likely due to the rise in FSH. During the luteal phase, the glucose levels can therefore be higher, and the diabetics might need to add more insulin. How strong the reaction is of course very individual, so it is important to track it yourself.

Another important hormone that regulates hunger is leptin. Leptin is produced in the fat cells and regulates our hunger by sending signals to the hypothalamus. It is not an immediate reaction but rather a long-term one, helping us to self-regulate our weight.[197] If we put on weight above our normal equilibrium, more leptin is produced, and we automatically eat a little less. If we lose weight below our normal weight, the leptin levels decrease, and we eat more. However, people with weight problems develop a resistance to leptin, so they have consistently high levels, but due to the resistance the natural regulation that makes you stop eating when you are full, is no longer functioning properly. Like many other hormones, leptin also changes cyclically.[198] Leptin increases over the menstrual cycle with a peaks mid-cycle together with the LH peak.[199]

Box of chocolates

I am a happy, and mostly functional, chocoholic, and I am not alone. About 30% of all women claim to have chocolate cravings. Many researchers have tried to figure out exactly why chocolate cravings are so common. It is, however, difficult to isolate what lies behind the chocolate cravings, as chocolate contains 380 different chemicals. One of the main suspects is anandamide, which is a chemical that acts the same way as THC, the chemical that makes you high when smoking marijuana. Anandamide triggers the neurotransmitter dopamine that makes you feel happy and also stimulates the learning, so that you remember that chocolate makes you happy. However, to reach the same effect as smoking a joint, you would have to eat 30 kg of chocolate.[200]

Another suspect to why we like it so much is phenylethylamine, a chemical related to amphetamines. Like amphetamines, this chemical causes blood pressure and blood-sugar levels to rise, which makes you feel more alert and happier. Phenylethylamine has been called the 'love drug' because it quickens your pulse, as if you are in love.[201]

During the luteal phase – and during pregnancy, which has a high level of progesterone – women tend to eat more sweet food.[202] Just before menses, we also tend to consume more carbohydrates and fatty food. Especially women with severe PMS consume more carbohydrates in this phase, and there is a correlation between bad mood and carbohydrate rich food intake. Even though the causation is not quite clear, researchers have established that eating meals with a lot of sugar and little protein can actually alleviate symptoms such as depression, anger, sadness, and confusion in PMS. And yes, there was a control group to try out for any placebo effect, so it is not only psychological.[203] Chocolate cravings are the most common craving in women, a yearning that increases in the luteal phase.[204]

The blinded-box

You have probably heard of the placebo effect. This is when your brain is convinced that a fake treatment is the real thing, and your body will react positively and heal without being given any active medical substance. You can say that it is a healing through the mind. It is a powerful effect but both a blessing and a curse. It is a blessing when my mother could heal all my childhood illnesses with ice-cream and chocolate sauce served with a convincing talk. But it is a curse when you want to prove the effect of a real treatment. This is why in clinical trials you sometimes need to have a control arm where you give one group the medication you would like to test and the other group something that looks just like the real thing.

This is called a blinded trial. To make sure that the doctors do not act differently when they give the real versus the fake treatment, even the doctors are not aware of what is what. It is then called a double-blinded trial.

Digestion – a question of muscle?

Did you know that period diarrhoea is a real thing? And that it is very common to be constipated in the luteal phase? One of the many places where you find hormonal receptors is the smooth muscles of the gastrointestinal tract. Smooth muscles are also called involuntary muscles and form a part of the supporting tissue of blood vessels and internal organs, such as the intestines, stomach, and bladder. Progesterone binds to these receptors and relaxes the smooth muscles. As a result, high levels of progesterone can cause constipation as the muscle is a bit too relaxed to help the food pass;[205] it can also cause bloating. About 75% of women report that their bowel movements are linked to the menstrual cycle.

If progesterone relaxes the smooth muscles, the prostaglandins released during menstruation do the opposite. They cause smooth muscle to contract. They mainly do so in the uterus, but they also influence the surrounding tissues and can create muscle contractions of the intestines and bowels. This can in some women lead to diarrhoea and bowel pain.[206]

It is not quite clear how hormones influence conditions such as irritable bowel syndrome (IBS) but it is very likely that it does, since IBS is 1.5 to 3 times more common in women than in men.[207]

Gut-box

The latest research is showing more and more clearly how our gut influences our entire life and is a key player in all health aspects, somatic or emotional. The gut microbiome – i.e., the sum of all the gazillion microorganisms, mainly bacteria[208], living in our gut – has been shown to influence mood, allergies, cardiovascular disease and much more. The microbiome is a rather new field of research which is why there are still many unknown issues.

But here again, hormones play their part, with oestrogen influencing the composition of the microbiota.[209]

Alcohol – how much of the good is good?

Even though you might be tempted to drink more alcohol in some phases of the menstrual cycle than in others, there is no evidence that alcohol is digested differently in different phases, and hormones only play an indirect role in how women react to alcohol. Since oestrogen changes the body fat composition and makes women store more fat compared to men, women have a harder time getting rid of alcohol. Indeed, alcohol is stored in fat, whereas it is dispersed in water. There is also another factor, which seems to have a larger impact than the difference in fat, that makes women more vulnerable to alcohol than men. It is an enzyme called alcohol dehydrogenase that is released from the liver that breaks down alcohol. Men have more of that enzyme, which helps them digest the alcohol when they consume it through the normal route, that is by drinking it. When alcohol is injected directly into the veins, the difference between men and women is much smaller.[210]

If the response to alcohol does not change much over the menstrual cycle, it can have an impact on women suffering from PMDD. It has been proven that drinking just before menses could have a degrading effect on the wellbeing of this population.[211]

Alcohol consumption in women seems to be linked to the actual production of oestrogen: drinking more alcohol increases oestrogen levels.[212] Higher oestrogen can have both negative and positive effects: the negative is that it slightly increases the risk of breast cancer,[213] and the positive is that it might slightly delay menopause.[214]

Body – why women have big butts

"You have the perfect body; you were just born in the wrong century." It was a sweet way to tell me that just because my *rondeurs* were not fully appreciated by our society, it did not actually mean that I did not look good. I was in my early twenties and laughed at the witticism (and still think about it now and then when I am in a phase when I feel that I do not conform to the modern beauty ideals). But what should a woman look like, anyway? Depending on time and place, the aesthetic canons vary greatly, and we appear biologically in all different shapes and variations. Who knows where evolution will bring us eventually? If biology had kept being the main driver of evolution, we would probably have developed kangaroo pouches in the future as this would have been useful for many purposes.[215] Firstly, because we would not have to give birth to babies that are way too large for our pelvis, and secondly because it is terribly convenient to always carry your baby around (and would also provide a handy place where to put our keys, smartphones and wallet). However, as modern advancements, such as assisted conceptions and births, tend to override biological needs this will probably never happen.

My son wanted me to call this book: *Why Women Have Big Butts*. I chose not to follow his unique and colourful advice, but I will – although not all women have big butts – explain why *on average* many of us do. Many of the traits associated with the feminine shapes are driven by oestrogen and progesterone. Thanks to these hormones we have breasts, larger thighs and yes, 'big booty'. If given the right hormones, men could also develop breasts and even breastfeed.[216] Oestrogen stimulates the breasts to develop and grow, it creates the duct system inside the breasts and makes sure that the fat we put on in puberty ends up around the hips and thighs. Women have generally a much higher percentage of fat than men. This is probably motivated by having additional energy storage for providing food for our offspring. The fat storage around hips and thighs is especially beneficial for feeding babies as it is providing essential fatty acids, which are important for the growing brains of the developing baby.[217] This fat is much less harmful than the fat around the belly. During puberty, the oestrogen helps to slow down the growth and increases sensitivity to insulin. Insulin influences the amount of body fat and lean muscle a person can develop.[218]

When my husband and I tried to get pregnant with our first child, my mum said to me: "You have got to be fertile, I was super fertile!" It was quite a frustrating statement, and I would add that if there is one single thing all humans born before 1978 (when the first IVF baby was born) have in common, it is that they have fertile parents. But my mother insisted: "I am sure that you are fertile, look at your broad hips!" Also, not a very helpful statement. It is a common belief that having broad hips is linked to higher fertility but there is no indication that hip:waist ratio would influence fertility in any way or be a sign of it. Some studies have correlated large breasts and narrow waist with a higher level of oestrogen,[219] a finding other studies were unable to confirm.[220] The currently most popular evolutionary theory explaining why women have developed fat deposits on their hips and breasts is that it would be a way for nature to signal that they have enough energy for reproduction.[221] Then again, male preferences regarding the female body appear to be more linked to cultural and individual preferences, than to any possible evolutionary advantages of choosing women with fat around the hips.

Breasts – a tangible change

My breasts always hurt a bit a few days before my next menstruation. This, of course, has to do with the changing hormones. Oestrogen stimulates the growth of the breasts and causes the breast ducts to enlarge, whereas the progesterone production causes the milk glands to swell. Both these happenings can cause your breasts to feel sore. During pregnancy, this change is even more manifest. The increased oestrogen and progesterone during the luteal phase of the menstrual cycle stimulate a rapid growth in cells, which is then balanced through a natural cell death in other phases of the cycle. This cell growth can lead to as much as 15% increase in breast size during the luteal phase.[222]

Breast-box

Cyclic breast pain, or cyclic mastalgia, normally occurs in the luteal phase as a result of the increased water content in the breast due to the increased hormone levels.[223] A recent study claims that 11% of all women have moderate to severe pain and about 58% mild discomfort.[224] Cyclic breast pain is not an indication of an underlying issue so unless it is perturbing your life, you do not need any special treatment. It seems to be more common in the perimenopausal phase, which is a period of cycle irregularity, and it disappears again after menopause. It is also more common in women who have breastfed, which is why it is suggested that breastfeeding causes changes in the breast that can later lead to pain.[225]

Bones – the scaffolding you should care about

Do not underestimate the importance of your bones. We take them for granted when we are young but if they have not mineralized properly during our youth and reproductive years it will come back to haunt us when we are old. Our bones are the scaffold that stabilizes our entire body and when it easily breaks, it is not only painful but a real nuisance for months and maybe years to come. My mum has terrible bones; every time she falls, she breaks something. She has broken her hip, arm and wrist several times already. It does not even have to be a dramatic fall, sometimes you have the impression that she just sat down a bit faster than usual, and snap, a bone is broken.

My mum is not unique: when they get older, many women suffer from frail bones and are very likely to develop osteoporosis, a bone disease which name literally means 'porous bones'. During their lifetime, up to 50% of women will break a bone – be it a shoulder, forearm, hip or spine – due to frail bones.[226]

Most people think that the adult skeleton is a stable structure of bone, but the reality could not be further from the truth. A bone is a living tissue, and hormones play an important role in its development. There are four major factors in bone modelling and remodelling: hormones and the charge we subject the bones to are the most important. Nutrition and concentrations of calcium in the blood play a key role as well.[227] The skeleton is remodelled continuously and is completely renewed every ten years.[228]

Testosterone stimulates bone building, whereas oestrogen prevents bone loss.[229] After menopause, when the oestrogen level is very low, bone loss occurs because more bone is resorbed than built. We talk about osteoporosis when bone loss has attained a certain level.

When the natural menstrual cycle is interrupted, this leads to less overall oestrogen which has a strong effect on bone quality. Already after 6 months of interrupted menstrual cycles, the bone quality of a young woman is equivalent to that of a woman over 50.[230]

To keep your bones in good condition you need to continuously load them. This is best done through physical exercise. If the bones are not loaded, bone mass will rapidly decrease, as can be experienced after space travel and a prolonged bed rest.[231]

Space-box

The gravity of earth has a natural influence on the load that we are experiencing on our bones, and it is something we take for granted. Astronauts on the other hand lose around 1–2% of their bone mass every month spent in space, because the bones no longer need to support the weight of the body due to gravity.[232] It might not matter much when you fly to the moon: the first Apollo missions lasted from a few hours up to 14 days. It does matter though for the long duration stay at the International Space Station that may last 3 to 6 months, not to mention what will happen when we fly to Mars. These long stays in space have resulted in losses of bone mass of up to 20% and highly increased fracture risk. Special exercise programs have been put in place on the space station, but this is not enough to entirely counteract the bone loss but together with supplements of vitamin D, calcium and some medicine it can at least be minimized.

Lung function – more sensitive than we think

Looking around at my friends and their children, I have started to think that boys had weaker lungs than girls. Although it is true for children, it is not for adults. After puberty, the pattern between the incidence of asthma in boys versus girls is suddenly reversed and once adults, the prevalence of asthma is nearly 50% higher in women than in men.

Asthma tends to decrease after menopause, but if you take hormone replacement therapy following menopause, the risk of asthma increases, even in non-smokers, which suggests that sex hormones may play a role in the development and progression of asthma.[233] Asthma symptoms can both become better or worse during pregnancy which makes the exact roles of the sex hormones in the disease even more confusing.

It is a well-known fact that the severity of asthma fluctuates over the course of the menstrual cycle.[234] For most asthmatic women, the best lung function is during menstruation when all hormonal levels are low. However, for another 20–40% of women with asthma, the symptoms worsen during menstruation.[235] This is due to the prostaglandins that are released during menstruation, a metabolite known to make asthma worse.

Women are also more sensitive to the negative impacts of smoking. For the same number of cigarettes, women are 20–70% more likely to develop lung cancer than men.[236] Some researchers have hypothesized that female sex hormone contributes to additional oxidative stress and greater airway injury.[237]

Heart – a temporary protection

A man suddenly grabs his chest, he makes a face, stumbles and collapses: we all recognize the typical signs of a heart attack as they are usually portrayed in many movies. It is a male heart attack; heart attacks in women look quite different. As with men, women's most common heart attack symptom is chest pain and/or discomfort. But women are somewhat more likely than men to experience other symptoms, particularly shortness of breath, nausea or vomiting, and back or jaw pain.[238] Although heart attacks are commonly associated with men, heart disease is the number one killer in women as well. In the US, one in five women dies of a heart disease.[239] What makes the matter more complicated is that a heart attack in a woman is more subtle than in a man. Instead of the 'classic' heart attack symptoms such as chest pain and tingling in the arm, they experienced anxiety, sleep disturbances and unusual fatigue.[240] Unfortunately, even doctors seem unaware of the differences between male and female symptoms. A 2018 study even showed that female patients seeing a female doctor were much more likely to survive than those who saw a male doctor.[241]

Yet women do have a natural protection from heart diseases during their reproductive years thanks to both oestrogen and progesterone. Oestrogen has many different effects on the cardiovascular system. It relaxes, smooths, and dilates blood vessels so that blood flow increases. Through these mechanisms, oestrogen limits harmful changes of the heart.[242] On the other hand, oestrogen can promote blood clots which can be very dangerous.

Heart-box

With age the shape of the heart changes. Whereas men's hearts grow, women's hearts tend to shrink after menopause. To compensate for the smaller volume, the heart needs to beat faster. This can have negative consequences and increase women's mortality.

Another effect of oestrogen is that it increases the good cholesterol and reduces the bad cholesterol. Cholesterol is a 'waxy' substance found in all cells; it is carried around with the bloodstream. Increased 'bad' cholesterol level is a significant risk factor for heart diseases in both sexes, but it has a significantly stronger effect in men compared to women. The 'good' cholesterol is beneficial because it helps remove other forms of cholesterol from the bloodstream. Higher levels of good cholesterol are therefore associated with a lower risk of heart disease. High levels of bad cholesterol can build plaques that attach to the inner walls of the blood vessels and make them narrower. Clots can then form and get stacked in the narrowed space, which might lead to a heart attack or a stroke. Women have higher levels of the 'good' cholesterol and a lower cholesterol synthesis than men throughout adult life. After menopause this changes, and postmenopausal women have higher levels of 'bad' cholesterol than premeno-pausal women or men of the same age.[243] However, even with these bad odds, women seem to have lower risk for repeating and severe cardiac disease.[244]

Sports – can we optimize for the cycle?

I go jogging two or three times a week, although I do not picture myself as being athletic in any way. It is simply a way of staying in shape so that I am fit enough to do the things I really enjoy, which is, to be completely honest, not running. And again, I have the impression that I can feel a difference depending on where I am in the menstrual cycle. In my follicular phase, I feel much more powerful than in my luteal phase. Having researched the topic further and tried to find confirmation of this theory, I started doubting my impressions. Indeed, most studies have been unable to uncover any link between athletic performances and the menstrual cycle. There are, however, a few factors that could affect how we perform[245] and could explain why, for instance, the female US soccer team claims to optimize their training after the menstrual cycle[246] and why I feel those differences at my more modest level.

One thing that has been shown is that the menstrual cycle influences injury prevalence. The risk of injuring the ligaments in the knee are much higher in the high oestrogen phase, just before ovulation.[247] Water retention, temperature regulation and fuel metabolism also change with the menstrual cycle. We know that the temperature is higher in the luteal phase when progesterone is high. This is also linked to delays in sweating, which could create issues when exercising, especially in warm temperatures. However, in female elite athletes it could not be shown that they had problems with temperature regulation over the menstrual cycle.[248]

Other studies claim that muscular training has a larger impact during the first two weeks of the menstrual cycle.[249] It remains unclear whether hormones have a significant effect on muscle strength during the cycle, but there is evidence that oestrogen improves muscle quality – if not its size – while it is being built up. We know that oestrogen generally has an effect of promoting cell-growth and it also improves the function of myosin, a protein that is active in the muscle contraction.[250] The impact becomes more significant in menopause where women lose a lot of muscle strength with the drop in oestrogen.

The painful-box

Pain tolerance is influenced by the balance between progesterone and oestrogen. We are the most sensitive to pain during the late luteal phase just before menstruation and the least sensitive just before ovulation when oestrogen peaks.[251] Oestrogen modulates pain in a similar way to how it makes us happier, it increases the amount of serotonin and increases the number of opioid receptors.[252] Despite rumours (mainly spread by men in my near proximity) that men are more sensitive to pain than women, no significant difference in pain tolerance between men and women has been shown.

Pulse, temperature and breathing rate increase in the luteal phase, factors that could impact/influence physical performances over the cycle.[253] It's not clear exactly how these changes occur, progesterone might act directly on the hypothalamus to regulate the temperature, but this has not yet been proven in humans.[254] Oestrogen, on the other hand, lowers the temperature so the relation between progesterone and oestrogen is what drives the temperature change, not the absolute amount of progesterone.

Progesterone is also instrumental in the relaxation of smooth muscles. Smooth muscles are the muscles you cannot voluntarily control such as those in the bladder, uterus, intestines, stomach and arteries and veins. Most well-known is the relaxation effect on the smooth muscle of the uterus which is necessary to maintain the pregnancy and prevent early delivery. It has not been shown in humans yet, but progesterone also relaxes the bladder which scientists have proven in pigs.[255]

If hormones have such an impact on how the body reacts to training – what happens if you are on hormonal birth control? The Department of Sport Science, at Nottingham Trent University in the UK, looked at 42 studies investigating this and observed that the different studies all reported different results. In the cases where a difference was seen, it was subtle. Their conclusion was that all decisions with regards to hormonal birth control should be taken on an individual level to see what the best fit for each athlete was.[256]

Skin – so warm and soft

One of the reasons that might explain the theory of women being more attractive in their fertile phase is how the skin changes under the influence of oestrogen. There are both oestrogen and progesterone receptors in the skin, but the effect of oestrogen is the best understood. As in many other areas of the body, the oestrogen boosts the vascularization of the tissue. Therefore, the blood flow varies with the menstrual phase, increasing during phases with high oestrogen.[257] This makes the skin smoother and softer in the fertile phase. It can also make women's skin warmer.[258] The increased vascularization is also why women tend to bleed more in the event of a cut, but also why oestrogen can speed up wound healing.[259] During pregnancy it is the increased oestrogen level that lies behind the 'pregnancy glow' some women experience, with a 'rosy', beautiful skin.[260] Oestrogen is not only important for the vascularization in skin, it also impacts sweat glands, oil glands and skin pigmentation.

Just before menstruation, when the oestrogen level is low, the skin is thinner and the skin barrier permeability therefore higher, which makes it more suscep- tible to allergens and other irritants. Many skin conditions – such as dermatitis, psoriasis, and acne[261] – worsen in this period. On the other hand, other inflam- matory skin diseases are improved during pregnancy, when you have continu- ously high levels of oestrogen.[262]

Following the same process, oestrogen also has a role in preventing the skin from aging, by averting wrinkles.[263] Indeed, this hormone stimulates the produc- tion of collagen, a protein that binds tissues, leading to thicker and more moistu- rized skin.[264] This has an impact both during the menstrual cycle and even more so after menopause, when the oestrogen levels drop drastically.

The oestrogen deficit makes the skin thinner and dryer although it's not the whole story: natural aging and environmental factors and especially sunlight and smoking play important roles in destroying the collagen content.[265] The best thing you can do for your skin is to avoid the sun or use sunscreen, stop smoking and keep up a good blood flow through exercise.[266]

Having acne and oily skin is mainly driven by the amount of androgens. This could explain why some women get acne just around ovulation, as this coincides with a peak in testosterone, but in general, acne is most common just before menstruation. The reasons are less clear but – since oestrogen has a controlling effect on the oiliness of the skin – when its levels suddenly drop, the skin becomes oilier and could facilitate the breakthrough of acne.[267]

Vision – clear or blurry?

When I was pregnant, I went to get new glasses only to be kicked out from the optician's office. They told me to come back when I had stopped breastfeeding and that buying new glasses now would only be a waste of money. If I was very surprised, I admired their honesty. I had never heard that pregnancy influences the vision in any way, but apparently it does.

The various hormones affect the visual system in different ways. Since there are oestrogen receptors present in many different parts of the eye, it is not surprising that varying levels of oestrogen impacts the eye and the vision in particular.[267] One of the mechanisms is that oestrogen makes the cornea less stiff. Progesterone too could play a role because, as we mentioned earlier, it is known to relax smooth muscle and there are smooth muscle cells in the eyes contracting the lens. The hormonal levels can affect the oil glands in the eyes, which can make the eyes more or less dry. The dryness and change in refraction can lead to blurriness.

The major hormonal transitions – puberty, pregnancy, and menopause – lead to larger changes in vision. The changes that happen during pregnancy are reversible, which is why the optician advised me to wait. The vision during pregnancy can become blurrier, and you might experience some problems focusing. The changes around puberty and menopause are not reversible. During menopause you become less near-sighted and during puberty more near-sighted.

Face-box

The fact that the vision changes along the menstrual cycle is not the reason why we are attracted to different faces in different phases of the cycle.[268] Someone actually bothered to check.

Voice – why we are not baritones

Considering the strong difference between male and female voices, it is not surprising that hormones also play a role there. Oestrogen makes the voice box smaller and the vocal cords shorter. This gives women a high-pitched voice compared to men. Just like with the vision, the major transitions of the voice happen together with the major hormonal transitions – puberty and menopause – but to some extent also over the menstrual cycle. Studies have shown that the subjective evaluation of the voice changes over the cycle, even though this could not be quantified with acoustic measurement techniques.[269] The changes in voice over the menstrual cycle do seem to be very subtle in general, but hormones still appear to have some impact on the vocal cords.[270]

During pregnancy there are major changes in the voice of which some are directly linked to the hormones and others indirectly. The indirect changes are linked to the changes in posture due to the pregnancy and the reduced lung volume when the baby is pushing the diaphragm. The more direct impact of the hormones is the swelling and the dilatation of the blood vessels of the vocal cords. Interestingly, many of the changes become stronger after the baby is delivered and influences the voice for up to a year after delivery with the voice becoming lower-pitched and more monotonous.[271]

Immune system – the stronger sex

While editing this in early 2020, a third of the world's population – myself and my family included – were locked up and confined to our homes due to the coronavirus that was spreading over all continents with an amazing speed. The virus resembles flu, but it strikes much harder on the respiratory system. Though the mortality rate of the virus is relatively low, its virulence is excessive; the elderly population has been more severely hit by the virus. At the time I am writing this, the most affected countries are China, Italy and South Korea. In all these countries, it has been observed that the mortality rate is much higher in men than in women.[272, 273] To some extent it might be the lifestyle, but there is more to the story.

Women have a very different immune system compared to men. Sometimes it serves us well, and sometimes it turns against us. Even though I have refused to believe it for a long time, my husband might be right, when he tries to convince me that a 'man cold' is a real thing, and that it is normal for a man to be completely miserable for a week with a cold that a woman would barely notice. It is indeed true that men in general get worse from a virus, and a woman's immune system will beat down a virus more efficiently.

When the body gets exposed to injuries, infections and toxins, it reacts by triggering an inflammation. An inflammation is not the same as an infection. Infection occurs when disease-causing invaders such as viruses and bacteria enter the body and start spreading whereas inflammation is the response of the body to that invasion.[274] Women have a more robust immune system, because it triggers a stronger inflammation as a response to invaders, such as cold viruses. Women are also less likely to get diseases like malaria, HIV and influenza, and when they do, the disease has a lower intensity. However, a more effective immune system is not always an advantage. Sometimes an excessive inflammatory response is triggered in women which can lead to other problems. During the swine flu[275] pandemic in Canada, women had more than double the risk of death than men.[276]

The differences of the immune system between men and women can be linked to different factors, both to the genes and to the hormones. Many genes related to the immune system can be found on the X chromosome. Women have two X chromosomes whereas men only have one. One theory is that the female body can choose the X chromosome with the best immune system by selectively inactivating some of the genes bound to the other X chromosome.[277] Men on the other hand are stuck with whatever is on their one X chromosome. As a general rule, testosterone has a suppressive effect on the immune system, while oestrogen has an enhancing effect on the immune system.[278]

Eunuch-box

Some scientists even claim that men have a shorter life expectancy than women because of testosterone. Documents about eunuchs in South Korea from the mid-sixteenth to the mid-nineteenth century show that they lived almost two decades longer than their uncastrated peers.[279] Whether testosterone is the determining factor is not entirely clear so please do not castrate your male life partner yet.

Progesterone and oestrogen have a different impact on the immune system.[280] Whereas oestrogen increases the histamine and serotonin release, progesterone inhibits this release.[281] This makes us more sensitive to allergies during the fertile days just before ovulation, when oestrogen spikes.[282]

Progesterone lowers the response of the immune system by suppressing antibody production. This is necessary in order to allow for the foetus to implant in the endometrium during the luteal phase.[284] As a side effect, we are more sensitive to infections in the luteal phase, when we have more progesterone, but by some incredible trick of nature we tend to be more disgusted by germs in this phase, which helps us avoid getting sick.[285]

Not only does women's immune system react more strongly to infections but also to vaccines. Women consistently report more frequent and severe local and systemic reactions to viral and bacterial vaccines than men. It is why it is imperative to integrate sex as a biological variable in clinical trials. The clinical trials of a herpes vaccine give a powerful example. When the vaccine was tested on a mixed group of men and women in 2002, no protection from infection was observed in the mixed group. When data were analysed by sex, the efficacy of the vaccine was of 73% in women and only 11% in men: vaccine was able to provide protection against the development of symptoms for genital herpes in women but not in men. Unfortunately for the women, the vaccine never made it to the market.[286]

The effect of a strong immune system can sometimes be confusing. When we feel sick, it is often due to the reaction of the immune system, i.e., the inflammation that is triggered to exterminate the virus. Our immune system is set off faster and more often than the one of men, so we might feel sick more often, but we will conquer the virus faster and have a higher rate of survival than men. A backlash of a strong immune system is that it can also be turned against us, which is why women are more susceptible to autoimmune disease.[287] Almost 80% of all patients with autoimmune diseases are women. As this difference in prevalence of autoimmune disease between men and women only becomes important after puberty, it is believed that oestrogen and progesterone play an important role. The exact development of autoimmune disease is not yet known – it could also be linked to abnormalities in the X chromosome – but, nevertheless, sex hormones play a central role.

Autoimmune-box

The women-to-men ratio is the most pronounced for Sjögren's Syndrome, Hashimoto's thyroiditis, lupus, rheumatoid arthritis and multiple sclerosis (MS). For all those diseases it is six to nine times more likely for a woman to get the disease.[288]

For sexually transmitted diseases, the cervical mucus (the fluid in your vagina that is produced by the cervix) plays an important role in immune protection. The texture, composition and the antimicrobial properties of the cervical mucus is heavily influenced by the hormones. Studies of the cervical mucus suggest that it changes significantly between the menstrual phases and that we are more resistant to infections during the ovulatory phase.[289] In the case of HIV, the quality of the mucus plays an important role in the risk of getting infected by the virus.[290] Certain synthetic progestins, types of artificial progesterone that are used for hormonal contraception, can lower the immune system. For instance, scientists have found that one type of progestin injection, used for birth control, significantly increases the risk of HIV,[291] partly by modifying the mucus.

Immune-box

The effects of hormones vary depending on the autoimmune diseases. In MS, the immune system attacks the myelin sheaths surrounding the nerves. Women are about three times more likely to get MS, but this is to some extent counterintuitive, since oestrogen actually has both anti-inflammatory and neuroprotective effects in MS. In lupus, the immune system attacks normal healthy tissue and organs. Women are ten times more likely to get it, and here oestrogen is not protecting but making it worse.[292] In this case, progesterone seems to have different effects depending on its concentration. At a lower level, it seems to stimulate the occurrence of lupus, whereas at a higher level, like in pregnancy, it seems to have a positive effect.[293]

As high progesterone lowers the immune response and prevent hyper-inflammation, pregnancy (which is marked by very high progesterone levels) can help against pain caused by rheumatoid arthritis.[294] The same effect can be seen in MS where symptoms can be better during pregnancy. However, in asthma, symptoms can both become better or worse during pregnancy.[295]

Autoimmune diseases are examples of inflammatory disease, but there are many other inflammatory diseases that are not autoimmune, such as asthma or the inflammatory bowel disease. An inflammatory disease is when the immune system triggers inflammation over a longer time due, for instance, to an allergic reaction. Inflammatory diseases are all more common in women and also more severe. The severity often changes at puberty, during the menstrual cycle, and after menopause.[296]

Cancers – do the hormones matter?

The common denominator of many cancers affecting women are not hormones, but rather other factors such as lifestyle, environment, age and family history.[297] The gene mutation BRCA1 – a genetic example of these factors – became notorious when the famous actress Angelina Jolie decide to have both her breasts and ovaries removed in 2013. She had tested positive for the mutated BRCA1 gene and wanted to reduce her risks of developing the disease. Having this particular mutation gave her a risk of 87% to develop breast cancer and a 50% of developing ovary cancer. The reason why Jolie got tested in the first place was that her mother died at a young age and that family history is the main factor that will predict your risk.[298]

Hormones are not the main factor which influence cancer risk, but they can play a minor role. The risk of breast cancer slightly increases with more oestrogen exposure. This matter is often discussed in relation to hormonal replacement therapy that can be used to relieve symptoms related to menopause. We will look deeper into this in the chapter on menopause. The link between oestrogen and breast cancer before menopause is more complex. Pregnancy is a period when oestrogen levels are high but pregnancies, and also breastfeeding, generally reduces your risk of breast cancer. Unless, you already have a small tumour: in that case, the additional oestrogen will promote the growth of the tumour.[299]

Phyto-oestrogens – the plant-based oestrogen that is strongly present in soy – on the other hand seem to have a protective effect on breast cancer which might sound a bit contradictory since the body's oestrogen has the opposite effect. As soy is a frequent ingredient in Asian diets, some speculate this could be the reason why Asian women have less breast cancers than others.[300]

When oestrogen exposure is much higher than normal, it does seem to increase the risk of cancer. Such an occurrence happens to women with a very high BMI, higher than 30, after menopause. As the fat cells produce more oestrogen, this increases the risk of developing breast cancer.[301] Another effect of a high BMI is the chronic inflammation that is caused by the increased amount of fat cells. As many as one in five cancers are believed to be linked to chronic inflammation.[302]

Since inflammation is such a strong factor in developing cancer, one of the hypotheses on why taking hormonal contraception protects against some cancers such as endometrial cancer[303] and ovarian cancer[304] is that it stops the inflammatory processes that are linked to ovulation and menstruation.[305]

Testosterone – lust for life

When I was in my twenties, I had a huge lust for adventures, and I was seeking excitement and kicks wherever I could get it. In my early twenties, it was mainly through partying and relations to friends and boyfriends, in my mid-twenties through travelling, and later through high adrenaline sports such as climbing, backcountry skiing and kitesurfing. In my thirties, I started calming down, the main reason probably being having kids, which completely changes your life, both on a logistical level and from a hormonal perspective. Now that I am in my early forties, my kids are more independent, and the excuse of the hormonal storms of pregnancy and postpartum are behind me. My life could include big adventures again, but I do not seek them as I used to. I still love the same sports, but the way I practice them is completely different. When you push your limits in the mountains, you often get into trouble and the triggered adrenaline rushes would keep me high for days. If I get myself into trouble nowadays, it just makes me miserable and I do everything I can to avoid it.

Why did I change so much? The reason probably lies in the testosterone. Testosterone is not an exclusive male turf: it is a critical hormone for women as well, stimulating fertility, sex drive, red blood cell production, muscle mass and fat distribution. Testosterone has been frequently linked to risky decision making. Both men and women with high testosterone levels show riskier behaviour than those with low testosterone and the effect is even more pronounced in women.[306,307] The testosterone concentration of a woman of 40 was found to be about half that of a woman of 21 which is why we all seem to calm down and become significantly more risk averse with age.[308]

T-box

Testosterone is a hormone we normally associate with men.
Not surprisingly, as the name comes from Ernest Laqueur in 1935
who isolated the hormone from bull testes and it literally means
'the hormone from the testicles'. Laqueur was not alone in his fascination
of this part of the male body, testis preparations had been consumed
since the mid-1800s to enhance virility, and testes from animals have been
transplanted to men.[309]

The name testosterone is a bit misleading as it also can be fabricated
outside the testes, like in the adrenal gland and in the ovaries in women.
Testosterone will push development of a lot of typical male traits such
as body hair, more muscles and stronger bones. In men, the testosterone
production takes off properly during puberty and is the main driver
of the bodily changes they go through in this phase. The sex organs grow
with a factor of eight before they turn 20. Testosterone also thins
the hair on the scalp. Having hair on the head does not mean a lack of
testosterone but a man without testosterone will never turn bald.[310]
Just like testosterone is not purely male, there are other hormones with
names from their actions in women that play a role in male sexuality,
such as follicle stimulating hormone (FSH) and luteinizing hormone (LH).
In men, LH stimulates production of testosterone in the testicles

and FSH interacts with testosterone to boost the sperm production. In their late forties or fifties, men's testosterone production will decrease, and they will also experience the equivalent of menopause, the male climacteric is around 20 years later than that of women.

In women, testosterone also varies with the menstrual cycle. Once the cycles have become regular, there is an increase of testosterone during the follicular phase that culminates during ovulation.[311, 312] The effects of testosterone on libido that we discussed earlier, could potentially be due to the mood enhancing effect it has in women. Some oral contraceptives increase the amount of sex binding globulins which binds testosterone. This decreases the amount of free testosterone in the blood. It could be a reason some women have a lower libido and less frequent orgasms, when they are on oral contraception. Again, the individual response in women varies a lot so it is difficult to make any generalizations.[313]

In men, testosterone is mainly produced in the testicles. In women testosterone is produced in the ovaries, adrenal glands, fat cells and skin cells. Normally, the testosterone is rather quickly transformed into oestrogen, but if too much testosterone is produced, the body can't keep up with the transformation which leads to an excess of testosterone, and consequently to problems such as PCOS, acne, facial hair growth, etc.

Having too little testosterone is also a problem, which can lead to a decreased sex drive, vaginal dryness, osteoporosis, etc. It can, however, be difficult to diagnose as the symptoms sometimes resemble depression. It is controversial whether adding testosterone actually helps women, research is still needed.

To PROCREATE OR NOT PROCREATE ...

When women are young, they spend an awful lot of time worrying about avoiding pregnancy. Once ready to start a family, it is all too common to suddenly realize that it was not that easy after all. Female fertility has been a mystery for a long time, and up until the 30s, not even doctors knew that it was only during a certain phase in the menstrual cycle one could get pregnant. One of the first studies about fertile days was done during the First World War, in 1916, when German soldiers were allowed to go home to their fiancées for only one day.[314] It was then possible to match how many of these visits ended up in a pregnancy and link it to the menstrual cycle. The study had many limitations – there were for instance no guarantees that the soldier was the actual father and not the postman – but it did bring to light a higher chance of pregnancy around mid-cycle.

In the 60s, the researchers thought the number of fertile days was around 3 days. In the 80s they changed their minds and approximated it to be around 10 days. More recent studies now estimate the window to be around 6 days, with variation depending on the individual women.[315]

When my husband and I decided that I should try to become pregnant with our first baby, I got a glimpse of how frustrating a trying-to-conceive journey can be. As the months kept passing by with no baby on the way, I got more and more nervous, and it did not take long until our entire life revolved around this one project. Once you enter that state of mind, things get even more complicated. There is no bigger turn-off than coming to your husband with sad, hollow eyes saying: "We must have sex now." Luckily, it worked out eventually, and after a bit more than a year of trying we expected our daughter.

One year might seem long, but it is still considered normal. Looking back, with the knowledge I have today, I believe several factors delayed the process despite being fertile. 'Mistimed intercourse' – a nice expression for having sex outside of the fertile window – could have been one reason. Also, I had just gotten off the pill and it took a while before my cycles became regular again. As soon as I had a 28-day cycle and ovulated around day 15, I got pregnant.

With our second child, I got pregnant before we even began to try. We faced other challenges though. In Switzerland, we get a first ultrasound 8 weeks into the pregnancy. As I considered the ultrasound to be just a quick formality, I went to the first consultation without my husband and with my 10-months old toddler on my arm. Of course, she could not stay still so, while getting a vaginal ultrasound, I tried to calm her down: the doctor completely freaked out and started screaming at me.

"KEEP HER QUIET!" she yelled.

As I tend to mimic the people I talk to, I shouted back at her: "I'M TRYING,

BUT I AM SITTING IN A GYNCHAIR WITH A PROBE UP MY ... WHY ARE YOU
SCREAMING ANYWAY?"

Still into the screaming game she answered: *"SOMETHING IS WRONG!!!"*

The foetus had no heartbeat and was never going to develop into a baby.
I had to go home and come back a week later for a second appointment. That
week I did not feel well. My body continued to be pregnant, but I knew that the
embryo was dead. It was a big turmoil of emotions: I was sad and yet secretly
relieved since closely spaced pregnancies are very stressful. Our mourning process was helped with the knowledge of how common early miscarriages are and
with the presence of our cute and wild toddler. The week after, my body had still
not understood that the foetus was not alive so after confirming a second time
with ultrasound the doctor gave me a pill to end the pregnancy. I went home and
bled for a bit over two weeks.

What happened to me is called to have a missed abortion, or a silent miscarriage. The pill I had gotten was an abortion pill. I have forgotten the name of the
pill but the most common one these days is Misoprostol which helps the uterus
contract and shred the lining to rid itself of the embryo.[316] Three months after
this loss, I was pregnant again with our son. Needless to say, I chose another
doctor to follow me through that pregnancy.

Our story is nothing compared to what many couples need to go through to
fulfil their dream of a family. Knowing your body well is not a panacea – for
getting pregnant or accepting the delays or failures – but, in many cases, it is
constructive and tangible first step in making this journey a little easier.

Missing-boy-box

During times of crisis, such as after earthquakes, natural disasters and even the 9/11 attack less boys are born. The theory is that the already pregnant women miscarriage easier and as the male foetuses are more sensitive, they are the first to go. On the upside is that the ones who get born have a better survival-rate in their early years. [317]

It is no wonder that it sometimes takes time to get pregnant. When you hear about all the things that need to be aligned for a successful pregnancy, it feels like a miracle that it ever works. Even though nature has managed to figure it all out, many couples still struggle with infertility. In some cases, it is possible to find a cause for infertility, such as issues with the menstrual cycle, the sperm quality, or mistimed sexual intercourse. Sometimes, the cause simply remains unknown. Is it one of nature's quality checks that has failed, is it simply bad luck, or is it something completely different? About 30% of infertile couples fall into the category of 'unexplained infertility'.[318] To be diagnosed as infertile, you should have had unprotected intercourse for 12 cycles, or 6 cycles if you are above 35, and still not pregnant. The reason why the criteria differ for older couple is that you need to act earlier, as fertility unfortunately decreases with age.

Fertility – what you need to know

As you have gathered by now, the female reproductive tract regulates fertility and having a fully functional menstrual cycle is a prerequisite to be fertile. The vagina, uterus and fallopian tubes change with the hormones over the cycle to provide an environment helping the sperm reach the egg so that they can fuse and later implant in the uterus.

The risky journey of a sperm

Fecundation used to be portrayed by doctors as the epic and brave journey of valiant sperms on their way to meet with some passive egg (a narrative bearing many resemblances with Rapunzel and other tales).[319] This myth has been debunked, and the impregnation is nowadays more accurately described. You will find that the real story is that Rapunzel and the other princesses are guiding the way for the prince and handing him all the necessary tools, to make sure that the journey is as smooth and easy as possible and leave them little risk of failure.

Corkscrew-box

The first person to ever look at sperm under a microscope was Antoine van Leeuwenhoek in 1677. He extracted the sperm by making love to his wife and, immediately after orgasm, taking a sample to put it under the microscope. This was a major scientific milestone,[320] although he got many things wrong about the nature of sperm, starting with the notion that everything needed for the development of a full person was contained in it, with the mother being just the mere bearer of this new life. He also described how the sperm swims by beating its tail back and forth. Scientists have revealed the inaccuracy of this conception this year only, in 2020, with the help of a high-speed camera. It captured the exact motion of sperms: they are in fact spinning their tail in one direction while rotating their head at the same time, much like a corkscrew. This movement propels them forward. Understanding the true motion patterns of sperm can have important future implications on IVF treatments.[321]

Now, back to sperm. Sperms are rather erratic little creatures and not terribly efficient in their swimming. It is therefore crucial that the quality of the sperm is good enough, meaning that it contains enough energetic swimmers in the ejaculation to allow for big losses on the way. That is, however, not enough. They need all the help they can get, or they would simply get lost, so the female reproductive system has a busy time helping them reach their goal.

Tadpole-box

Did you ever wonder why sperm look like tadpoles? This is most likely due to an adaptation to the environment. Just like tadpoles, the sperm need to move quickly through water. The sperm is a cell and looks like a cell, hence its round shape. But it still needs to navigate through an aquatic environment and, to do that, a little tail is the most helpful tool.

As the sperm first enters the vagina, it needs to reach the uterus. During the female fertile phase, the cervix opens, and a specific type of nourishing cervical mucus is produced by the lower regions of the uterus. The mucus selects the healthiest sperm and capacitates it as it swims through. The immune system in the uterus also plays an important role in this phase and manages to detect if the male sperm is of appropriate quality and a good match.[322] If not, it will not be allowed to pass.

The capacitation process – which can also be done in vitro during IVF – is a reaction where the mucus changes the surface of the sperm and hyperactivates it, as if the princess were giving energy drinks and power bars to the prince. Hyperactivation means that the swimming pattern becomes much more intense. This is necessary for the sperm to bind to the surface membrane of the egg and to start the reaction that will make the sperm penetrate and fuse with the egg. The sperm binds to the membrane for about a minute and thereafter penetrates it. This will, however, only happen if the egg considers the sperm fit, which is decided by another reaction on the membrane itself. You can see that there are many screening processes that it needs to go through in order for the female reproductive system deems it fit for the job.

Even the capacity of the sperm to find the egg may not be entirely random. There is some evidence of sperm-egg communication and the egg can even choose which sperm it would like to succeed to make sure it is the most genetically compatible sperm.[323] So, in this more accurate version of the fecundation, the egg gets to know her 'suitors', to communicate with them, to help them, to screen them and to choose 'the one'. We are definitely getting away from the sleeping beauty scenario.

The sperm-egg fusion happens inside the fallopian tubes, which has also done its fair share to help the sperm reach its goal by modifying its environment to guide the sperm in the right direction.[324] Drawing on the fairy tale analogy, the suitor doesn't cross by himself a dark and dangerous forest: the princess has created a path along the trees and put out torches so he will not lose his way. Once the sperm and egg have fused, it takes 5 to 6 days for the developing egg to reach the uterus, where the uterine lining will make another quality check – the third one – before it allows it to implant and start growing. By using sensitive pregnancy tests, the total rate of pregnancy loss before implantation was 46%, and 30% after implantation.[325]

After ovulation, the lifespan of the egg is very short. If not fertilized, it is estimated that it only lives approximately 24 hours after leaving the ovaries. This means that the sperm needs to be in the reproductive system, already waiting when ovulation happens. Sperm can stay alive for 5 days in the right environment, even though it is claimed that it is only fit for duty for 72 hours. Therefore, there are only around 6 days that intercourse can make you pregnant every month. The exact number of days depends on the quality of the cervical mucus, which is often linked to age, but in general it is between 2.5 and 7.5 days.

Fertility and age – don't be deceived

You will always have a friend telling you that she had no problem getting pregnant in her forties. It does happen, but it is very important to keep in mind that fertility strongly declines with age. Sometimes it is easy to believe otherwise thanks to all the celebrities you see pushing strollers around way into their forties. What you do not see is all the costly fertility treatments they have gone through. Our best reproductive years are in our twenties and fertility drops during our thirties. A healthy, 30-year-old woman has a 20% chance of getting pregnant every cycle. When she is 40, that number has dropped to 5%.[326]

Even though we might have healthy cycles well into our forties, other less visible factors weigh in. With age, both the quality of the eggs as well as the number of days of high-quality cervical mucus declines. The lower egg quality will lead to more chromosomal changes which leads mostly to miscarriages and sometimes to Down's syndrome. Not only does it become more difficult to get pregnant with age, but the pregnancy also becomes riskier with higher occurrences of high blood pressure, pregnancy diabetes and low birth weight. Therefore, it is important to get help as early as in your mid-thirties if you are struggling to get pregnant.

As we get older, the body pumps out more follicle stimulating hormones to make the best possible use of what is left of our egg reserve. This sometimes leads to more than one dominant follicle being selected, which is why twin births are more common in older women. The start of this stronger decline in fertility generally happens around 35 years of age but, as everything else, it is very individual matter. Even egg freezing has a lower success rate if the donor is over 35.

Getting pregnant – what can you really influence?

So, what can you do to get pregnant? You will probably find tons of recommendations on the internet, but most of them are not supported by science. A friend of mine swears by 'the bike'. 'The bike' is a position when you are lying on your back after intercourse, lifting your pelvis cycling with your legs in the air. She claims that this position helped all women in her family – including herself three times – to get pregnant. Although I am not denying that it could be a fun thing to do, there is so far no one has managed to prove that that any specific position during or after intercourse would impact fertility.

What is recommended to boost your chances of getting pregnant is frequent intercourse during the fertile window, that is during the 6 days leading up to and including the day of ovulation. By frequent, scientists mean once every 1 or 2 days, so there is no need to do it five times a day (unless this is what you want anyway).[327] It can be difficult to pinpoint the fertile days exactly.[328] It is also recommended not to smoke or to drink more than two drinks a day, not to use recreational drugs, and avoid most commercially available lubricants as they might mess with the natural mucus.

Control-box

Choosing the sex of your baby could be a whim or a matter of life or death, depending on the context. At the risk of disappointing you, scientifically: there is nothing you can do to influence it. Your chances are close to 50/50, with a slightly higher probability of getting a boy. The exact percentage of getting a boy or a girl depends on environmental factors, so it is not possible to set a fixed number.[329]

There is a popular idea that the sex of the baby can be influenced by the timing of intercourse. Sex is determined by which chromosome the sperm is carrying. The misconception that you can chose sex is based on the fact that sperm with a Y chromosome (male) swims faster whereas sperm with a X chromosome (female) lives longer.
The idea is then that if you have sex closer to ovulation, later in the fertile window, the chances of having a boy is higher, and intercourse earlier in the fertile window would lead to a girl. The basis of the reasoning is correct, but there is no clinical evidence that it works.

It is also claimed that you can influence sex with food. By eating food that changes the pH of your cervical mucus, the mother would favour one sex over the other. Also, an interesting theory but with no backing in the scientific literature.

One thing you should do if you are planning to get pregnant is to take folate or folic acid. This does not only increase your chances of getting pregnant,[330] it also helps prevent birth defects linked to the neural tube and heart valve.[331] A neural tube defect in the baby will prevent the brain from developing properly and, if the baby survives, it will be heavily disabled. Since the introduction of guidelines on taking folic acid during pregnancy, the incidence of neural tube defects has gone from 1–2 per 1000 babies born to under 1. It is also linked to malnutrition and the difference is stronger in countries where this is more frequent.[332]

A common belief is that stress influences your fertility negatively. "Just relax and it will happen," is something many struggling couples hear but there is no evidence supporting that. As counterintuitive as it might sound, your fertility will not be influenced by emotional distress.[333]

Light-box

There is a seasonality of birth rates. In Sweden, where winters are very dark, there is a strong fluctuation in the number of babies born dependent on the month. There is a birth peak in spring, with babies conceived during the summer months. In the southern hemisphere, those peaks occur with a 6-month shift.[335] This is most likely due to the more extreme changes in light that you have closer to the poles. Temperature has been compensated for in the study, so that is most likely not the cause. As all Swedes and other northerners know, you often have less energy in winter and, since the function of the ovary is also linked to the energy level, this may also play a role.

Obviously, birth rates do not only depend on female fertility: males play an equally important role, and the sperm quality is also linked to seasonality.[336]

Contraception – an old invention in need of renewal

For several millennia, humans have been imaginative in finding ways of controlling their fertility. Scriptures from ancient Egypt (1500 BC) mentioned the issue, so did other texts from the Roman empire and the old Greeks around year zero but it would take another 2000 years until we came up with reliable methods. Some of the early methods were reasonable but ineffective, such as wiping out the vagina after intercourse. Others had some effect, like using barrier methods. Other methods were arbitrary, and completely ineffective, such as the woman holding her breath at the time of ejaculation or jumping backward seven times after having sex.[337] The ancient Romans used animal bladders to protect women against contraction of venereal disease.[338] The first documented use of a 'condom' in Europe was in the mid-sixteenth century by the same man who named the Fallopian tubes after himself, the anatomist Gabriele Falloppio.

At that time, the primary use of the condom was not to protect against pregnancy but syphilis. Apparently, it worked well. Falloppio were very convinced about the use, he wrote that he had provided more than 1000 soldiers with a small linen cap drenched in a salty and herbal solution, to be pulled over the penis, and that not one single user had contracted the disease. Despite that, it was only one century later that penis sheaths fashioned of sheep intestines started being used to avoid pregnancy.[339]

Being able to control our fertility is a most amazing and liberating progress in the lives of women. Not being constantly pregnant has made it possible for women to access education and have more liberty in how to shape their lives.[340] But the big revolution only came with the pill in the 1960s and you will read more about that later.

The barrier methods for both men and women are still popular but today there are many options for controlling your fertility. You can do it naturally by simply abstaining during the days you are fertile (but make sure you get it right!). There are intrauterine devices (IUDs) you can implant in the uterus, contraceptive patches, and of course oral contraception ('the pill'). Each method has its own pros and cons, which I will not detail here. I will go through their main principle of action and show in which way they influence the body.

Barrier methods – like the male and female condom, caps and the diaphragm – have close to no influence on your body as their only action is to physically prevent the sperm from reaching the egg. There are, however, other methods that have a larger impact.

Having an impact on your body is not synonymous with having a negative impact. Many methods have a bad reputation, especially hormonal methods, but since everyone is different, those methods can have varying impacts. There is therefore no general right and wrong. Furthermore, whatever means of contraception you chose, it does not have to be final. You can always change your mind in case it does not feel right for you.

Intrauterine devices (no need to recoil anymore)

When I was younger, I was explicitly warned not to get an intrauterine device (IUD) before I had had all the children I wanted. It was believed that the IUD – or coil – could harm your long-term fertility, especially if something went wrong during the insertion or the removal. With the development of the devices this is no longer the case, and one of the benefits with an IUD, is that fertility is quickly restored after it has been removed from the uterus. With a good insertion technique there is no increased risk of infertility after the removal.[341] That being said, all interventions have a risk and the main moment of risk for all types of IUDs is the insertion, where there is an increased risk of infection during the 20 following days.

Intrauterine devices or systems are small devices that are inserted into your uterus for a longer time period. The span can vary between three and ten years depending on the brand. They prevent pregnancy by modifying the environment in the reproductive system, either through hormones or through a 'foreign body effect'. The presence of a foreign body in the uterus creates an inflammation which is toxic to sperm and impairs their capacity to move. It prevents them from meeting the egg and fertilizing it.[342]

In a copper IUD, the side effect of the inflammatory reaction is that more prostaglandins are produced in the uterus. This may lead to heavier and more painful menstrual bleedings. A hormonal IUD has the opposite side effect, reducing the bleeding after 3 months of use and sometimes stopping it altogether. The presence of the artificial progesterone, the progestin, works against the prostaglandins and stabilizes the uterine lining which reduces menstruation. Except for this difference, the hormonal IUD acts in a similar manner as the copper IUD: it is toxic to sperm. The difference is that it is covered with an artificial progesterone, called levonorgestrel. This hormone thickens the cervical mucus, which prevents the sperm from entering the uterus, and it thins the uterine lining, which prevents implantation. Common for all IUDs is that their main effect is not to prevent ovulation.

Side effects are rare for the hormonal IUD. However, not having any period may be a concern for some women to the extent that they remove the IUD and choose another method of contraception. Sometimes the progestin reaches too high levels in the bloodstream, which can lead to some androgenic effects on the skin (acne and hair growth) or breast tenderness.

'Useless-word'-box

Killing of sperm is called spermicide and is a term that is frequently use in medical literature.

After my two children and because I wanted to keep my natural hormonal fluctuations, I chose the copper IUD as a method of contraception. Furthermore, I am not disciplined enough to go for a natural family planning method. This choice turned out to be very beneficial in my previous job. No, I was not a lap dancer but I was researching how to detect hormonal fluctuations over the menstrual cycle. This was done using physiological parameters such as temperature, pulse rate and breathing rate and the cupper IUD allowed me to use myself as a test subject. The side effects of the copper IUD are an increased menstrual bleeding, up to 55% more in fact, which, both for me and many other women, can lead to anaemia. Despite this negative effect, I am happy with my choice. I enjoy feeling the effects of the changing hormones.

Weird myths surrounded intrauterine devices, and the copper IUD in particular. For instance, the anti-abortion movement accused it of being a 'serial abortion device'. Their idea was that it was not preventing fertilization, but just preventing the implantation of the fertilized egg. It is a misconception: copper IUDs do prevent fertilization. As its etymology suggests – from English in the late nineteenth century, 'against + conception' – it is a true contraceptive device.[343]

Oral contraception – the pill of empowerment?

Already in the 1930s researchers had seen that progesterone could prevent ovulation in animals but it would not be until 1960 that the first commercial oral contraception was available. When the pill first appeared, it was marketed for cycle regulation because it was illegal both in the US and in Canada to have a public discussion about contraception. This was considered both obscene and morally corrupt and the pill could only be prescribed to married women. In Canada it only became legal to prescribe the pill for contraception in 1969. Around that time the pope also issued a statement in the name of the Catholic Church that condemned the pill as an 'artificial' means of birth control and, thus, as sinful. Ironically, the first clinical investigation of the pill was driven by two Catholics, Dr John Rock and women's health advocate Margaret Sanger, and took place in Boston in 1950. Due to the prohibitive laws in the US it was complicated to conduct trials on contraception, so another early pill researcher, Dr Gregory Pincus, decided to move his trial to Puerto Rico. The lack of regulations unfortunately went together with lack of respect for the 200 women who participated in the trial. They were given no information about the dangers of the trial and when they started reporting side effects such as nausea, dizziness, headaches and blot clots, their testimonials were neglected.[344] Towards the end of the 60s, concern about side effects of the pill were taken seriously and today's pills have only a fraction of the levels of progestin and oestrogen compared to the early versions.

Despite the bumpy road of the early days of the pill, there is no doubt that the introduction of the pill in the 60s has been the main driver of women's economic empowerment,[345] improving their access to education and better paid jobs. Not to mention the huge benefits in women's health. In developing countries, the access to oral contraception has cut the number of maternal deaths by 40% over the past 20 years, merely by reducing the number of unintended pregnancies. A Lancet article has even posited that another 30% of maternal deaths could be avoided by extending the access to contraception for all women.[346]

That being said, the use of oral contraceptives is still not harmless for all women. You often hear that men would never accept doing to their bodies, what we are doing to ours and would never accept the side effects that we accept. There is room for improvement, and it is important to keep researching new and better ways to prevent pregnancies.

Some women on the pill do not suffer from any physical or mental side effects, whereas others react very strongly to it. Biology being so complex – and the number of factors involved so high – it is impossible for anyone to predict how you will react. The pill uses the natural feedback system of our endocrine system to prevent ovulation by mimicking the hormonal state of the luteal phase or early pregnancy. This might sound like a thing that is not so good for women but remember that previously women were pregnant most of the time, so it is a 'natural' state.

The main active part of the pill is progestin. There are many different types of progestins but all of them are artificial versions of progesterone. You cannot use normal progesterone because it breaks down too quickly in the body. Progestin, just like progesterone, inhibits the luteinizing hormones, which would normally trigger ovulation.[347] It is precisely what happens during pregnancy. The levels of progestins being much lower than the ones occurring naturally in the luteal phase, the endometrium is not built up as well as it would during a natural cycle which grants an additional protection against pregnancy. If, against all odds, the ovulation was to take place anyway and the egg be fertilized, the endometrium would not be thick enough to allow the egg to implant. The progestin also produces a cervical mucus that is thick and difficult for sperm to penetrate.

Pro-box

Progestin is very similar to the progesterone that it mimics, and it binds in the same way to progesterone receptors. The difference between the artificial versions and the natural version is how it attaches to other receptors as well.[348] In earlier versions of the pill especially, the progestin would bind to androgen receptors and trigger a reaction which could lead to acne breakout and the development of male traits such as being verbally less fluent but better in spatial ability.[349] At the time indeed, most of the progestins were extracted from testosterone. Even today, the pills with more androgenic properties tend to be the cheaper ones, and they have been shown to influence the brain. There are many different pills which all uses different progestins on the market and each of them has different effects, which is why it is important to find the one that is most adequate for your body.

The most common pill nowadays is a combination pill that also contains oestrogen. The oestrogen used is the same in most pills and very close to natural oestrogen. The oestrogen prevents the follicles from developing by blocking follicle stimulating hormone and prevent breakthrough bleeding. A breakthrough bleeding happens when the endometrium is not stable enough and thus is shed, in the same way as during the normal menstruation. Oestrogen stimulates the growth of the cells in the endometrium, stabilizing it. Oestrogen also increases the effect of the progestin, which makes it possible to lower the

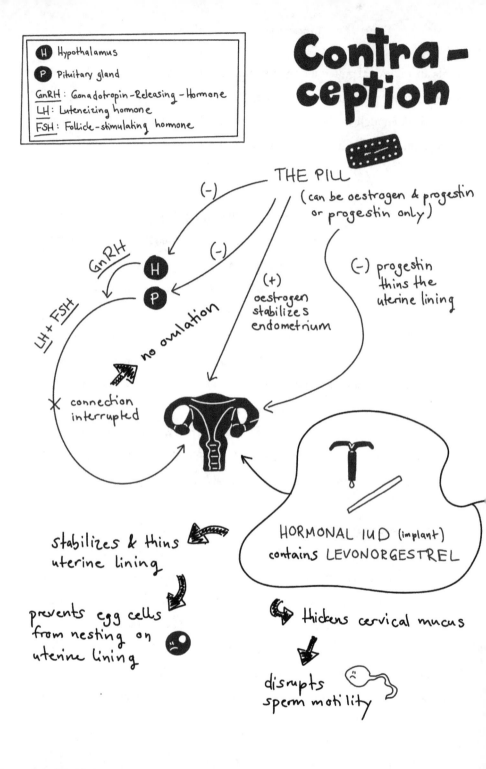

progestin dose.[350] The dosage of oestrogen is very important and should not be too much. One of the most dangerous side effects of the pill is thrombosis, a problem linked to oestrogen.

Oestro-box

In the 1920s, researchers found that you could extract oestrogen from the human placenta and umbilical cord and use for therapeutic purposes. They later found that they could extract it also from urine of pregnant women. When the researchers discovered that similar oestrogens were present in the urine of pregnant horses, this opened the door to commercialization as this could be extracted in much larger quantities. In 1941 it entered the market under the name of Premarin, a name that hints the origin from the pregnant mare.

When the hormonal dose is high in the pill, there might be a delay in getting pregnant when stopping. This is the case with older pills, some of which are still on the market and used today. The newer low-dose hormonal contraception does not have this problem and 3 months after stopping you will have regained your fertility.[351]

Oestrogen is generally good for the arteries since it prevents them from the thickening and hardening due to fat depositions. However, in the veins, oestrogen increases the risk of blood clotting. Therefore, oestrogen therapy is good if it is applied directly in the start of menopause, before the arteries have become rigid and unresponsive. If you start taking oestrogen after a long period without it, the arteries might already be stiffened, and if you go back on oestrogen again, the risk of blood clots is increased. Blood clots, or thrombosis, is a well-known side effect of the pill. It is rare, but when it happens, it is very serious. The main risk factors for thrombosis are family history, weight, and smoking; women with a high risk should rather get the progestin-only pill since the oestrogen is here the hazardous hormone. The danger for young, non-smoking women is, however, very low. In the future, genetic testing could potentially check known thrombophilic factors, so that you could choose the right pill for you.[352]

Cancer-box

The pill will also protect you against other things: both the risk of endometrial cancer[353] and ovarian cancer[354] are significantly decreased by the use of combination pills with both oestrogen and progestin. This protection against some cancers has been explained by the hypothesis of 'incessant ovulation' and 'incessant menstruation'. Both ovulation and menstruation are inflammatory processes; since inflammation is linked to higher risk of cancer, preventing cyclic inflammation by suppressing ovulation and menstruation could be beneficial.[355] On the other hand, the risk of cervical and breast cancer increases slightly with the use of these kinds or pills, but since breast cancer is rare in younger women, that risk is not so heavy. To avoid cervical cancer, the best thing to do is to get the vaccination against human papilloma virus (HPV), which is the virus that causes it.[356]

My first month on the pill was an emotional rollercoaster. It was the summer after my first year at university, and I had found a student job. At work, I would inexplicably burst into tears and I run away to hide in the ladies' room (as this was an engineering job, it was a very quiet place in the 90s). It is hard to sort out the reasons for these emotional outbursts: were they linked to the pill, was it because I was young and full of feelings I could not control or was it a combination of the two. For the first time, I was away from my boyfriend who I was head over heels in love with, and this could have played an aggravating role too. Emotional state usually stems from a mixture of factors, but my experience was not unique. Up to half of all women that stopped using oral contraceptives name negative mood changes as the main reason.[357] However, since many women expect those kinds of negative side effects, it often becomes a negativity bias[358] or a self-fulfilling prophecy. As a matter of fact, since many women expect the pill to have negative effects on the mood, they become overly attentive which might make them more prone to 'notice' negative emotions. In my own case, the symptoms disappeared after the first month even though I remained very emotional. I would not blame the pill for that, as I have stayed that way even after quitting the pill.

Depressive-box?

A 2016 Danish study[359] looked at data from over one million young women (average age 24.4 years) who were given oral contraception for the first time. Years later, on average 6.4 years, the women were followed up and the researchers checked the women's use of antidepressants, if they had gotten other major psychiatric diagnosis, cancer, venous thrombosis, or needed infertility treatment. The study showed a correlation between oral contraception and an increased risk of depression, something which scared many people. The study was heavily criticized though, and its methodology questioned. One problem was the difference between correlation and causation. If you remember from the introduction, this is when two observations are linked but it is not clear whether one is causing the other. In this study, there was a correlation between depression and the pill but no proof that the depression was caused by the pill. One possible reason for the correlation could be that women already suffering from PMDD often get prescribed oral contraception to lessen the symptoms, and these women have a higher incidence of depression.

Despite the amazing things the pill has brought women, it is not the right solution for everyone. When you are on the pill, hormones are evenly administered over the whole cycle. This means that you will miss the positive mood-boosting peaks before ovulation, but you will also suffer much less from the downs in the premenstrual phase.[360] Based on the evidence currently available, it is, however, likely that taking hormonal contraception can lead to mood-related side effects, in particular for women with a history of depressive episodes. There are indications that the capacity of recognizing and reacting to emotions are altered. On the other hand, there are also reports of positive effects of hormonal contraception use on mood in some women, especially for symptoms of PMDD.[361] As you learned in the chapter on hormonal impact, PMDD is triggered by certain levels of progesterone. Altering those levels artificially through the pill can in some women alleviate the symptoms.

One way hormonal contraceptives might influence us is through the hormone oxytocin. Oxytocin is the hormone that makes us bond with other people, both partners and children. It has a particularly strong effect in new couples and in mother-child bonding. It also influences our sexuality and is known to increase the intensity of our orgasms and how much we enjoy sex in general.[362] There are some indications that the response to oxytocin is modified in women using hormonal contraceptives.[363] This mechanism might also be behind the observed effect that being on the pill can influence your choice of partner.[364]

The findings on how the pill influences libido are rather contradictory.[365] On the one hand, a known effect of the pill is an increased production of sex hormone binding globulin (SHBG), which binds testosterone. It leads to a diminution of free testosterone, which could potentially result in a drop of the libido. Sexual desire is, again, a very complicated matter, and testosterone not being the only player, it is very hard to draw simple conclusion on the sexual side effects of hormonal contraceptive. On the other hand, a 2019 article published in the International Journal of Reproductive Health assessed that the pill could boost the libido, on the principle that the fear about an unwanted pregnancy is gone.[366]

Stress-box

The hypothalamic-pituitary-adrenal (HPA) axis dealing with the stress response is permanently altered by the use of oral contraceptives. This changes how you react in stressful situations. The cortisol released during stressful events normally leads to better memorization which makes us super alert and allows us to better remember the situation we are in, so that we can learn from it for the future.[367] A consequence is that women on birth control become less good at recognizing anger and other feelings in people's faces, on average. This is a weakness normally considered a more masculine trait. However, do bear in mind that this effect is on average and, when digging deeper, some studies show that this problem affects mainly women with a certain genetic composition.[368] This is one of the reasons why you will always find women who love their pill and women who hate it.

One advantage of hormonal contraception methods is that if we want, we can entirely suppress menstruation, which can be – as many female readers have experienced first-hand and as male readers will have gathered by now – rather debilitating. This kind of contraception can also help against many issues such as dysmenorrhea, abnormal uterine bleeding, endometriosis, mittelschmerz, acne or undesired facial hair.[369] In fact, the pill is sometimes prescribed by doctors precisely to deal with these ailments. However, it is very important to keep in mind that if the pill has been given to ease a problem, it will not get to the root of the problem, but only cover the symptoms. Hence, when you will get off the pill, the issue will reappear. This is particularly important to remember when trying to get pregnant. In some cases, the underlying problem may prevent you from getting pregnant, so make sure you stop the pill in good time, to sort out any potential issue.

Should we be afraid of adding hormones to our lives? Some hormones can work in our favour and give us huge benefits. Even though we should stay vigilant and monitor undesirable side effects, we should not be overly anxious or suspicious. When starting on a hormonal birth control, a good option to keep track of potential side effects is that you make a diary before and after starting it to follow your reactions. Research is still missing on this topic, and the individual differences are so large from woman to the next that until we have truly personalized medicine, no one can tell you what the best solution is for you.

Emergency contraception

Did you ever experience the panic of a condom breaking? When I was younger, I somehow thought that unprotected sex would get me pregnant immediately. Knowing a little bit more about my menstrual cycle could have alleviated some of that stress. In my early twenties, I had no clue about such things, but luckily there were other solutions.

Emergency contraception – or the day-after pill – is a hormonal pill that prevents or delays your ovulation for the ongoing cycle. Normally, it contains progestin, levonorgestrel – the same as in the IUDs – and sometimes also oestrogen. It should be taken within 72 hours of unprotected intercourse, but the earlier the better. It decreases the risk of pregnancy by 94%.[370] As it delays ovulation, it is not the same thing as getting rid of a foetus. However, it is not a very pleasant way of contracepting since a very common side effect is severe nausea which is particularly tough because if you vomit, you need to retake the pills.

Dystopia-box, Future of only women?

Being able to reproduce without involving men might sound like science fiction but, in theory, we know how to do it. Researchers have already managed to use stem cells of a mouse to recreate the genetic material from sperm (and egg) needed to create a fertilized oocyte that they could implant in the mouse.[371] The 'surrogate mother' gave birth, but the offspring did not make to adult life. It is of course even more complex in humans, and we are far from filling the gap between science fiction and science.

Disruptions of the Menstrual Cycle

Menstrual health is about much more than fertility alone. Some claim that the menstrual cycle is our fifth vital sign, a strong indicator of our overall health. The four other vital signs are: body temperature, pulse rate, respiration rate and blood pressure. In the previous chapters, we have discovered how hormones released during a healthy menstrual cycle played a fundamental part in strengthening our body. They build up our bones and muscle strength, protect our heart, keep our skin young and much more. We do have a strong interest in making sure that menstrual cycle works as it should and if it does not, it is important to understand why. This chapter is about different ways this fine-tuned system can be disturbed.

If you are healthy and your life is in balance, your menstrual cycle will have a similar length every month, but when stressors from your daily life influence the cycle, it becomes more irregular. Such irregularities are mainly due to influences occurring during the follicular phase, the part of the cycle where the hormonal coordination is directed by the hypothalamus. The hypothalamus is the link between the endocrine system and the nervous system, and it plays an important role in many brain functions. It is therefore not surprising that the menstrual cycle is closely linked to our general health and wellbeing. If the hypothalamus receives signals that our body is not in a state where pregnancy is a sound idea, it will suppress ovulation, and hence the menstrual cycle, to spare our bodies of that burden.[372] The second half of the cycle – the luteal phase – is beyond the control of the hypothalamus, so this phase is normally more stable in its length and less influenced by external factors.

Interestingly, our bodies can handle very extreme conditions. Space travel, for instance, does not necessarily disrupt the menstrual cycle (even though many female astronauts chose to suppress bleeding by using hormonal contraception).[373] However, our lives here on earth seem to have a harder impact on our bodies due to the daily stress and strict diets we sometimes subject ourselves to.

Stress-box

In stressful situations, the hypothalamus decreases the release of gonadotropin-releasing hormone (GnRH), leading the anterior pituitary gland to reduce the amount of follicle stimulating hormone (FSH). As a result, it takes longer for the eggs to mature, which delays ovulation and hence prolongs the follicular phase.

Too much stress – or extreme physical activity not compensated with the appropriate amount of energy intake – is linked to slightly longer follicular phases (and therefore also cycles).

Amenorrhea is the medical term for the absence of menstruation. The most frequent causes of this ailment are hormonal imbalance caused by an underlying condition such as polycystic ovary syndrome (PCOS), hypo-thalamic amenorrhea or hyperprolactinemia. Thyroid disorders are less common but are also potential causes, and so is obesity and stress.

Another reason which we will all encounter sooner or later is repro-ductive ageing. One day, the egg reserve in the ovaries becomes too low to produce sufficient oestrogen to trigger ovulation and the menstrual cycle stops. This is menopause.[374]

Smoking-box

Smoking seems to boost the FSH production making the eggs mature faster. This shortens the follicular phase (and again, the cycle). The shorter menstrual cycles might contribute to smokers running out of eggs faster, and thus reaching menopause earlier.[375]

Hypothalamic amenorrhea – the balance of energy and stress

A lot of research on menstruation in Britain during the eighteenth century was based on patients from the lower social classes. It made it difficult to learn much about regular menstruation and understand the link between menstruation and fertility. Indeed, since many of the lower-class patients were poorly nourished – especially during the winter months when there was a general lack of food – they simply did not menstruate anymore, despite not being pregnant.[376]

These women suffered from something that is still common today, hypothalamic amenorrhea, an ailment occurring when menstruation stops due to the hormonal coordination of the hypothalamus. Causes can extend from extreme emotional stress to acute weight loss, chronic malnutrition, or extreme exercise. One of the key players in hypothalamic amenorrhea is leptin. Leptin is the hormone signalling that you have eaten enough. While you are eating, the leptin released from the small intestine has an immediate effect on your hunger: by hindering it, it makes you stop eating. If you have been eating more than usual for a longer time and produced more fat cells, the fat cells will produce more leptin to inhibit your hunger until you reach your normal weight again. To produce the hormones necessary to trigger ovulation, namely FSH and LH, you need sufficient amounts of leptin. It is impossible to maintain a regular, healthy menstrual cycle when levels of leptin are too low.[377] The lowest amount of body fat that is believed to be necessary for this is 22%. This is a higher percentage than most female athletes are estimated to have. Gymnastics, marathon running, and cycling are typical sports were the competitor's percentage of body fat on average is lower than this limit. In sports such as tennis and swimming it is more frequent that the practitioners have a percentage above this level. It is therefore no wonder that exercise is often mentioned as a reason for hypothalamic amenorrhea. In reality, it is rarely the only cause, but a combination

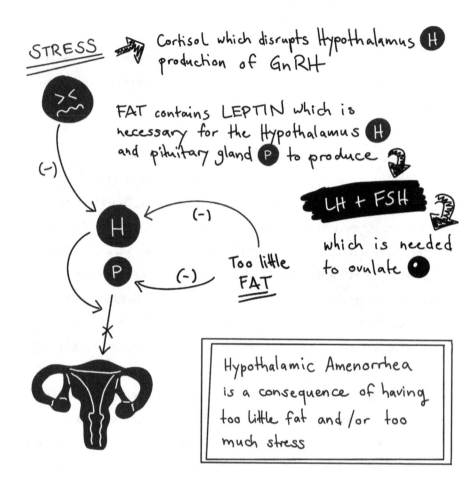

STRESS → Cortisol which disrupts Hypothalamus **H** production of GnRH

FAT contains LEPTIN which is necessary for the Hypothalamus **H** and pituitary gland **P** to produce

LH + FSH

which is needed to ovulate ●

(-)

H ← (-)

P ← (-) Too little **FAT**

Hypothalamic Amenorrhea is a consequence of having too little fat and /or too much stress

of physical activities and a low body weight such as displayed by some runners or dancers.[378] Illness and travelling – especially over time zones – can also cause the hypothalamus to shut off the menstrual cycle. As many as 3–5% of all women suffer from amenorrhea lasting three months or longer.[379]

Fat-box

Leptin is also believed to be the main driver of puberty. Therefore, it is necessary to have enough body fat to trigger the first menstruation. The critical amount for this is around 17% and to maintain the regular cycles you need 22%.

When a friend of mine, a passionate runner, lost her period during a longer timeframe, her doctor told her not to worry as long as she did not want to get pregnant and just enjoy not bleeding. This is not a point of view shared by all physicians. As a consequence of the interruption of the menstrual cycle, much lower amounts of oestrogen are produced, and oestrogen has many essential effects on our body and wellbeing. The most immediate effect of insufficient oestrogen is bone loss. The bone quality of a young woman who has had 6 months of diminished oestrogen is equivalent to that of a woman over 50 years old.[380] Oestrogen has also a very positive effect on the cardiovascular system. We know a lot about the effects of reduced oestrogen level in older women, but much less in younger women with amenorrhea. It is, however, a symptom that should be taken seriously. Researchers have shown that half of younger women with cardiovascular disease have low oestrogen. To prevent bone loss and other problems due to lack of sex hormones, oral contraceptives are usually prescribed. This solves some of the side effects of the hypothalamic amenorrhea but is not an answer to the underlying problem.

Another stress-box

When the brain is sensing stress, hypothalamus starts releasing a corticotropin-releasing hormone. This hormone tells both the pituitary gland and the adrenal gland to start producing cortisol. Cortisol is the primary stress hormone. It increases the glucose in the bloodstream, enhances your brain's use of glucose and increases the availability of substances that repair tissues. It also suspends all functions that are not necessary for our 'fight and flight' response, such as the digestive system and, you guessed it, the reproductive system. With cortisol, the whole menstrual system is hampered.[381]

The cortisol produced by the adrenal gland will also contribute to halting the ovulations. The cortisol tells the pituitary gland to stop producing luteinizing hormone, which is the hormone that triggers the ovulation. The cortisol also acts directly on the ovaries, where it can inhibit oestrogen and progesterone production.[382] Very high cortisol levels are widespread in most women suffering from amenorrhea.[383]

Psychological stress is a frequent cause of an interrupted cycle, and likewise an interrupted cycle can cause psychological stress. Women with hypothalamic amenorrhea have significantly higher depression scores, greater anxiety, and increased difficulty coping with daily stress as compared to healthy controls.[384]

Obesity – a matter of resistance?

If too little food may harm your menstrual cycle, too much of it will harm it as well. When you start accumulating too much fat, especially around the waist, the hormonal balance may be disturbed, which can lead to chronic inflammation of the body.

Fat-box

Fat cells around the waist produce oestrogen, leptin, and cytokines. Cytokines are signalling molecules, and the ones produced in the fat cells promote inflammation. An excessive amount of these cytokines causes chronic inflammation which is very harmful for the body and often leads to insulin resistance and diabetes.[385] Inflammation is the body's process of fighting against something that wants to harm it. When this process is constantly lingering, we talk about chronic inflammation.

Up to a certain point, being overweight is not a problem for your cycle and your fertility but when the body-mass-index (BMI) goes above 30, insulin resistance becomes very common, which can lead to hormonal imbalance, i.e., both too much male hormones and oestrogen. This can in turn harm the development of the follicles and even lead to polycystic ovaries.

Derailment-box

Insulin resistance means that the cells cannot use the insulin properly.[386] The body's demand for insulin then increases, and the response of the body is to create more of it. It triggers a chain reaction and the derailment of the whole system, to the detriment of fertility, among other effects. Extra insulin prompts the ovaries to produce more androgens (male hormones) and the liver to produce less sex hormone-binding globulin (SHBG). This means not only that more androgens are produced but also that less of it is bound, which leads to even more free androgens and to freer oestrogen. Increased levels of these hormones further disturb the production of FSH and luteinizing hormone LH.[387] This can harm the development of the follicles, and even lead to polycystic ovaries which we will discuss below.

Leptin produced in the fat cells can also influence the ovaries directly and has a deteriorative effect on the development of the oocytes. This in turn reduce the production of oestrogen and progesterone which can influence how well the uterine lining builds up which can make it more difficult for the oocyte to implant and hence even further harm fertility.[388]

Obesity leads to many issues with regards to the menstrual cycle and fertility, but the good news is that even a small weight loss can have a big impact on regaining ovulation.[389]

Polycystic ovary syndrome (PCOS) – an unfortunate chain of events
Having too much hair growth in the face and on other parts of the body, while the hair on the scalp gets thinner, sounds like every woman's worst nightmare. It is, however, some of the symptoms of one of the most

common hormonal disorder in women, namely polycystic ovary syndrome or PCOS. The hair growth comes from an excess of male hormones or androgens – so called hyperandrogenism – which can have many secondary effects such as hair loss on the scalp, more body hair as well as acne.[390]

The precise definition of PCOS is still a widely debated topic among specialists. Today, it is generally accepted clinical definitions include a combination of high levels of anti-Müllerian Hormone (AMH), multiple cysts on the ovaries, and irregular cycles. PCOS is the most common endocrine disorder in women and the most common reason for infertility with 6–10% of all women being affected and, depending on the exact definition, as many as 15%.[391]

In the previous section, it was described how chronic inflammation and insulin resistance can lead to an excess production of androgens, and how this can disturb the hormonal balance. When this hormonal imbalance inhibits the production of LH, the follicles that are growing inside the ovaries will never get the final push to release the egg. Instead, the follicles continue to grow and become fluid-filled sacs, referred to as cysts. Those cysts can become as large as 7 cm in diameter, but most commonly, women get many, at least a dozen, small follicles not large enough to qualify as cysts.

Catch 22-box

The growing follicles contain AMH and as the cysts are indeed growing follicles, the more cysts the more AMH. The high levels of AMH are most

Polycystic ovary syndrome

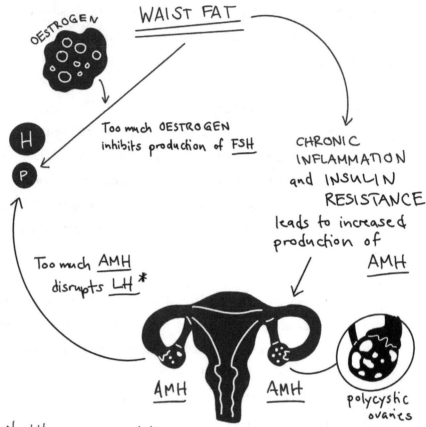

OESTROGEN

WAIST FAT

Too much OESTROGEN inhibits production of FSH

CHRONIC INFLAMMATION and INSULIN RESISTANCE leads to increased production of AMH

H

P

Too much AMH disrupts LH *

AMH

AMH

polycystic ovaries

* No LH means no ovulation, even though the follicles have grown. The follicles that didn't become eggs become cysts instead.

H Hypothalamus

P Pituitary gland

LH: Luteinizing hormone

FSH: Follicle-stimulating hormone

AMH: Anti-Müllerian hormone

> likely playing a role in inhibiting FSH. As FSH helps turn androgens to oestrogen this will lead to an excess of androgens.[392] This can become a negative spiral, the amount of androgens just get higher and higher[393] leading to more and more problems with ovulation, which in turn leads to more and more cysts.

Obesity is one of the causes for PCOS, with up to 80% of all women with PCOS having a BMI above 30 (the healthy range is considered being 18–25). Weight issues are not the only cause of PCOS: researchers have pointed to family history as reliable indicator, even though it is unclear whether it is a genetic or environmental matter, or a combination of both.[394] A team of scientists of the University of Lille, France, has recently found out that the exposure of the foetus to high levels of androgens – and in particular AMH – in the womb was a significant factor in developing PCOS.[395]

The close association of PCOS with obesity has led to many undiscovered cases among women with a BMI in the healthy range. In lean women as well, PCOS is believed to be a consequence of insulin resistance.[396] Up to 70% of women with PCOS – with weight problem or not – have insulin resistance, meaning that their cells cannot use insulin properly.[397] PCOS is also overrepresented in female athletes, who are normally very thin despite a lot of muscle mass. A professor in reproductive medicine from the Karolinska Institute in Stockholm has proposed that this could potentially be a natural selection, since women with PCOS have more androgens such as testosterone, they will develop more muscles which can boost their performance.[398]

Crochet-box

The journey of being diagnosed with reproductive issues is often a terrible rollercoaster for women. As with many other women, a friend of mine only realised that something was wrong with her hormones when she quit the pill to start trying to conceive. She suddenly got plenty of acne and she rather quickly got the diagnosis; PCOS. She was given hormones and to use her own words, the doctor 'burned' her ovaries. She most likely got a treatment called laparoscopic ovarian drilling (LOD). A laparoscopy is when you make a small incision in the abdomen wall and insert a thin tube with a high-intensity light and a camera. The camera sends images to a video monitor which allows to get high-resolution images from inside the body in real time, without open surgery. Ovarian drilling is a surgical technique where membranes around the ovaries are perforated using a laser beam.

This is not the most common treatment, when the main issue of the PCOS is infertility. The first line of care is normally to give clomiphene. Clomiphene is a drug that induces ovulation and works similar to oestrogen causing the egg to mature from the follicles and be released. Only if this medicine is not helping the next step of ovarian drilling is taken.[399] The intervention was successful, and she quickly got pregnant with her first child. She would have been very happy with this outcome if it were not for a new worry on the horizon. Due to the acne her PCOS had triggered, she had gone to a skin care specialist, who prescribed her vitamin A for her skin. The problem is that an excess of vitamin A in early pregnancy can cause birth defects if taken during the first 60 days of the pregnancy.[400] Luckily, she quickly realised it and ended the treatment and got away with a scare.

Her journey did not end there. For her next child she went to another doctor and got another laparoscopy. The doctor was pessimistic with regards to her chances of conceiving but gave her a crocheted handkerchief to help her get pregnant faster. She did what every woman should do if her doctor gives her a crocheted handkerchief for any other purpose than to wipe her face; she left and never came back.

With her next doctor and her third laparoscopy, she now found out that she had endometriosis. (You can read more about endometriosis in the chapter about the uterus.) Was there a link between her two conditions? There is currently no answer to that question.

In a recent study from 2019, researchers looked at a cohort of 240 women who underwent a LOD for PCOS treatment after the clomiphene medication had not worked. They found no increased incidence of endometriosis in these women compared to the incidence of asymptomatic endometriosis in the general population.[401]

To deal with the endometriosis she was thrown into yet another treatment round. This time she was put in an artificial menopause to stop her menstrual cycle so that ovaries would have time to heal. This sudden menopause caused horrendous mood swings, sweating and hot flashes. For nine months she had to endure that state until the treatment stopped, and she got her cycle back. Her efforts paid off with a happy ending – she immediately got pregnant with her second child.

Luteal insufficiency – a late consequence?

Sometimes, all the first steps of a successful conception go according to plan; ovulation takes place and the released egg fuses with the sperm. But then something goes wrong: the wall of the uterus is not strong enough to hold on to the embryo. This leads to an early miscarriage. In 35% of recurrent miscarriages, an insufficiently built uterine wall seems to be the cause.[402] This is referred to as a luteal insufficiency, since it is a consequence of the corpus luteum not producing sufficient amounts of progesterone in the luteal phase. Normally, such an insufficiency can be traced back to the follicular phase. If too little FSH is released the follicle from which the egg originates will not turn into a strong enough corpus luteum and this means that it is not able to produce the right amounts of progesterone.[403]

Apart from issues in the follicular phase, another potential reason for a low progesterone production is the lack of in cholesterol in the blood to produce the necessary hormones.[404] This can, in rare cases, be caused by a very strict vegan diet.

Luteal insufficiencies can be diagnosed through very short luteal phases and irregular bleedings.[405] There is, however, no consensus on what a short luteal length is, with some researchers suggesting 9 days and others less than 12 days.

Hyper/hypothyroidism

Issues with the thyroid can disturb the hormonal balance and with that your menstrual cycle. The thyroid gland is another endocrine gland, which is situated on the front side of our neck, just under the Adam's apple. Among other things, the thyroid gland regulates the metabolism. Two conditions related to this gland affect women much more frequently than men: they happen when the thyroid produces either too much, or two little hormones. One in eight women will develop thyroid problems during her lifetime.[406]

Boiling frog-box

You have probably heard the stories about what happens with frogs and boiling water. If you put a frog in hot water, it will jump out immediately. If you on the other hand put it in cold water and heat it up gradually, it will stay until it gets boiled. A similar thing happened to a friend of mine in her 30s. She had developed hypothyroidism but did not realise it until she got diagnosed. All the changes were creeping onto her, little by little. The fatigue got a tiny bit worse every day and so did her cognitive capacity. It became harder to remember the right words to use in sentences and it made her frustrated, but she never thought anything was wrong with her. The changes came so slowly that it felt natural. During a regular check-up, the doctor saw that her

> thyroid gland was very large and did a blood test which showed that
> her thyroid stimulating hormone (TSH) levels were too high, this is the
> sign of an underactive thyroid gland.

Too much or too little thyroid hormone can make your periods very light, heavy, or irregular. Thyroid diseases also can cause your periods to stop for several months or longer. If your body's immune system is the cause of thyroid disease, other glands, including your ovaries, may be involved. This can lead to early menopause (before age 40).[407]

Hyperthyroidism means that the thyroid gland produces too much hormone. The most common disorder associated with hyperthyroidism is an autoimmune disease, Grave's disease. The opposite condition is hypothyroidism, where the thyroid produces too little hormones. It is linked to Hashimoto's disease, another autoimmune disease.

The way the thyroid hormones affect the menstrual cycle is not well understood yet, but it could be either through a direct effect on the ovaries or by influencing the pituitary gland to inhibit the production of gonadotropins. Both too little and too much of thyroid hormones disrupts the menstrual cycle and the former, or hypothyroidism, leading to very frequent bleeding, and the latter, or hyperthyroidism, to almost no bleeding at all. However, to make things more confusing, it also happens that bleeding stops in hypothyroidism.[408]

Hyperprolactinemia

Hyperprolactinemia occurs when the pituitary gland produces too much prolactin, the hormone usually released to stimulate breastfeeding. The production of milk without being pregnant or breastfeeding is one of the main symptoms of hyperprolactinemia. Its causes can be hypothyroidism or benign tumours on the pituitary gland, but also the consumption of some medicines against pain (opiates), high blood pressure, depression or nausea, etc.[409] Prolactin inhibits the production of GnRH, which in turn decreases LH and FSH,[410] which prevents ovulation, and this is a chain of events that also happens during breastfeeding.

Endocrine disrupting chemicals

Bisphenol-A – those baby-bottles...

Around the time I had my first child in 2008, there was a lot of talk about Bisphenol-A (BPA) in baby bottles. BPA is a chemical used since 1960[411] to harden plastic and make it transparent, which makes it very popular, especially in bottles. The concern was that BPA mimics oestrogen and may leak into the baby formula, rendering the boys less fertile and pushing the girls into an earlier puberty. All parents in Switzerland, and probably elsewhere, were very careful not to buy anything containing BPA, but it would take until 2011 for the European Union to ban BPA in baby bottles and until 2012 for the US.

As it turned out, the problem was very real. Researchers showed that an early exposure to BPA could indeed hinder babies' neurological development. The same BPA is also suspected to play a role in children's obesity and the development of diabetes.[412] The full extent of the effect of these endocrine disrupting chemicals is still to be unravelled, and a lot of additional research is needed. It has now been shown beyond doubt that they do have an impact on chronic diseases such as obesity, diabetes mellitus, reproduction, thyroid, cancers, and neuroendocrine and neurodevelopmental functions.[413]

Baby bottle-box

The endocrine disrupting chemicals impact the reproductive system in many ways. BPA binds to oestrogen receptors[414] and, thereby, it influences the hypothalamus to secrete less GnRH. This directly affects the menstrual cycle by inhibiting follicular development and ovulation and reducing the quality of the oocytes. It does not only lead to hormonal disruptions but also to a structural change of the uterus and vagina.[415] Most organs have a fairly stable number of cells but in places such as the endometrium in the uterus and the lining of the vagina, the number of cells increases and decreases depending on the hormonal balance. Therefore, these organs are the most sensitive with regards to endocrine disruptors. Due to the same reasons, it is therefore the testicles that grow sperm that are the most sensitive parts of the male body.[416]

Phthalates – what is in your bathroom cupboard?

Phthalates – another group of chemicals – is a kind of chemical 'plasticizer' found in everything from personal care products to floors. They are extensively used in cosmetics, with personal care items containing phthalates including perfume, eye shadow, moisturizer, nail polish, liquid soap and hair spray. They have similar disrupting effects as BPA.

Hair dye and hair straightener are products that contain a lot of compounds known to be both endocrine disturbing and promote cancer. A recent study showed a strong increase in breast cancer risk with the use of hair dye, especially in black women.[417] Before raiding your bathroom and throwing all your personal care products though, keep in mind that phthalates are just one of the many chemicals that put you your health in jeopardy, and that the most important risk factors – such as gender and family history – are out of your control anyway.[418]

Drinking water – a matter of location

Containers and cosmetics are not the only sources of endocrine disruption and concern. There are also worrying reports about the amount of oestrogen and other hormones in the drinking water. The effects of the additional oestrogen are particularly strong in fish, where the males get more feminine traits and sometimes even develop ovaries.[420] The levels of oestrogen vary a lot depending on geographic location, places with particularly high amounts of oestrogen are regions with intense farming, such as the American mid-west and southern Australia. Other places with a high

level of oestrogen polluted rivers are countries close to the Mediterranean basin and Asia as well as some countries in South America. In places which have oestrogen pollution above the recommended limits it can be harmful for humans and impact both reproductive health and breast cancer risk.[421] It is a common belief that this pollution comes mainly from hormonal contraception, but a study has shown that only 1% of the oestrogen comes from that source.[422] Other sources are believed to be livestock and other pharmaceuticals.[423] Scientists struggle assessing the human health risk, causality being very complex. On an individual level, one of the things you can do is to check the levels of oestrogen in the tap water where you live. To this day, hormones levels in in European waters (apart from a few places close to the Mediterranean) are not too worrying; the situation on a global level and the impact on aquatic life around the globe is, however, quite worrisome.

Action-box

It is very difficult to avoid all the endocrine disrupting chemicals as they are present almost everywhere. What you can do then? You could try to avoid plastic containers for food and beverage[424] and look over your cosmetics products. Furthermore, you could try to protect yourself. There are indications that folates and soy can have a protective effect against BPA's and phthalates' influence on our bodies. Studies have shown that the decreased fertility linked to BPA and phthalates can be reversed if the women have an adequate intake of folates.[425, 426] Therefore, consuming folic acid can help maintain your fertility despite some BPA exposure. The same effect was also shown with soy products and a soy-rich food intake will diminish the impact of BPA.[427]

Soy-box

Phytoestrogens mimics the effects of the body's own oestrogen.[428] It is an oestrogen produced by plants and consumed through our diet. The most common phytoestrogen is soy, but wholegrain foods and seeds are also rich in phytoestrogen. There is a lively debate in the scientific community on the actual effects of soy in the diet, and its potential influence on the menstrual cycle. There is, however, no evidence of increased oestrogen in menstruating women with a normal consumption of soy. Due to some results in animal models, there have been rumours about soy causing breast cancer. Other studies have now debunked such effects, and no increased occurrence on breast cancer has been seen in humans. In Asian women, there are even indications that a soy rich diet prevents breast cancer, results that could not be reproduced with Western women.[429] One could suspect that adding phytoestrogens would somehow harm men's fertility, and some men were worried that they would develop breasts by eating too many soy products. Those myths have all been debunked, and it seems that no harm is done to sperm,[430] fertility,[431, 432] or testosterone levels in general.[433] And no, eating soy will also not give men boobs.[434] That being said, it is not yet clear what influences it could have when you feed your baby with soy-based formula. Some researchers see some influence but nothing that cause alarm but keep your eyes open for future research on this topic.[435]

A 'NORMAL' CYCLE

Cycle length and regularity

It is great to have a normal cycle. Sure, but how do you define 'normal'? As is often the case, answering this question is more complicated than one might think. In high school, I was taught that a normal cycle lasted 28 days. Now many recent publications suggest that the average length of a cycle is of 29 days. The range of what is considered normal is of course often far from the average.

Normal-box

It is quite comical – if not alarming – to witness the struggle to reach a consensus on the length of a 'normal' cycle. I have been sitting in meetings with experts on gynaecology from all over Europe and the US, seeing first-hand how they fail to come to an agreement. The propositions include all combinations from a lower limit of 21 to 24 days to an upper limit of 35 to 38 days: it is quite a range! A literature review presents the same inconsistencies. In 2007, a group of experts decided on a cycle span of 24 to 38 days,[436] but these guidelines have also been challenged. A paper analysing a menstrual app with approximately 600,000 menstrual records found out that the average cycle length was of 29.3 days with a standard deviation of 5.2 days. These findings mean that, if the data is normally distributed within this large population, 70% of the cycles lasts between 24 and 35 days.[437] The average length of a menstrual cycle might also vary by country: an Indian study examined over 8000 cycles and concluded that the most common cycle length was of 30 days.[438] Asian women have been reported to have longer cycles on average, with the follicular phase being longer than for Caucasian women. It is difficult to know if it is linked to diet or genetics,[439] and many of the differences between ethnicities could be explained by BMI.[440]

The regularity of the cycles is another indicator of what a normal, healthy cycle is. Then again, we enter a minefield where specialists strive – and fail – to reach an agreement. The most common definition is a variation between the shortest and longest cycle of 5 days, with one exception allowed per 6 months. In practise this would mean that for instance a variation between 27 and 32 days is considered regular, and if one cycle is longer or shorter than that every 6 months everything is still in order.

Regularity depends on your age. Cycles are the most regular from the mid-twenties to the end of the thirties. It is normal to have irregular cycles in the first couple of years after puberty also towards the end of our repro-ductive life, when we approach menopause in our late forties. As we grow older, the cycles tend to become shorter, the average cycle length being around 30 days at 25 years old and around 28 days at 35.[441] The cycles sud-denly get longer again the years preceding menopause.

Hormonal balance

As you must have gathered by now, the hormonal machinery follows a very subtle and intricate pattern. Like musicians in an orchestra, hormones must play their part at the right time with the right tune for the melody to by harmonious. In this context, hormonal balance is not some New Age concept, but the prerequisite for a healthy menstrual cycle. You need to have all the hormones varying at the right time at the right levels, meaning that there is enough follicle stimulating hormone to help the follicles mature and therefore start producing oestrogen. Luteinizing hormone should be triggered at the correct moment to release the egg during ovula-tion, and what is left of the follicle should turn into a strong enough corpus luteum to create progesterone in sufficient amounts to build up a nutritious uterine lining.

If the cycles are regular and with a cycle length and number of bleeding days within the 'normal' range it is a sign that that the hormones are beha-ving as they should. If you have regular cycles within the normal range, there is no reason to suspect that you are not ovulating. The cycles can still be ovulatory, even if they diverge from this pattern but if this happens often it is worth checking with an obstetrician/gynaecologist to exclude any underlying issues.

Progesterone offers an additional sign that you have ovulated. As the corpus luteum that produces the progesterone will only be formed after ovulation, there will be no progesterone without ovulation.[442] To get an indication that your progesterone levels are in order, it is not necessary to take a hormonal test. Indeed, one of side effect of progesterone is the alteration of many physiological parameters such as temperature, pulse rate and breathing rate.[443] By tracking one or several of those signs and observing a rise during the second half of your menstrual cycle, you would get a good idea on whether ovulation has taken place or not.

The easiest parameter to track is temperature. The temperature needs to be measured exactly at the same moment every day, before getting out of bed and with a thermometer with high sensitivity (two decimals). The change in temperature between the follicular and luteal phase is only expected to be around 0.2°C.

Oestrogen can have an opposite effect on temperature which is why it is common to see a dip in temperature just before it starts to rise again due to the progesterone.[444] This dip is exactly when oestrogen is the highest during the cycle.

It is, however, very important to notice that in spite of having a healthy menstrual cycle, not all women get the rise of temperature in the luteal phase, and even more women do not get the dip in temperature just before ovulation. If you have this pattern in your cycle, you can feel confident that you have a healthy cycle; but not having it does not mean that something is wrong.

Synchronized-box

Among the myths surrounding hormones and menstruation, the one about households of women synchronizing their menstrual cycle over time is quite widespread. In a famous study from 1971, scientists allegedly showed that college roommates were synchronizing their menstrual cycles.[445] Thereafter, many studies tried to repeat this finding, to no avail. Later publications claimed that the study was flawed, and a 2013 article reviewing several studies concluded that synchronizing cycles did most likely not exist.[446]

Synchronization with the moon is another popular myth. Studies were here again unable to verify this legend, even in indigenous populations – living far away from modern infrastructures and all their artificial lights.[447] From an evolutionary standpoint, such a synchronization would be hard to account for, and especially not as long as men's sex drive was not synchronized in a similar manner.

CREATING LIFE – THE LARGEST TRANSFORMATION

Pregnancy is the biggest transformation a woman's body will ever go through. The body will change and adapt to nourish and grow a small oocyte – less than a millimetre in diameter – into a complete human being. This chapter will focus on what happens in the body from conception until the new life, the baby, is born. One of the fascinating facts about pregnancy is that it is not your own body that is triggering this transformation. You can look at the new life as a foreign body that starts to grow inside you and will hack into your system and successively take control by releasing hormones into your blood. Those hormones are responsible for all the changes you will experience, not only during the 9 months of pregnancy but over the years following birth.

It is no wonder being pregnant is such an overwhelming experience, both physically and mentally. How women experience this can vary tremendously. Many women consider pregnancy to be the most magical adventure they ever went through, whereas some women see it as an ordeal, suffering continuously throughout the 9 months. Others hardly seem to notice that they are pregnant.

I have been pregnant twice, and it has been two incomparable experiences. During my first pregnancy, I was in perfect balance, happy and attentive to all the changes happening inside my body. I got the well-known 'pregnancy-glow'; when the skin gets rosy and makes the face appear fuller. I felt amazing and often got compliments, even from strangers. I stayed in great shape throughout the whole pregnancy, and the day before giving birth I went for a two-hour walk, ran up and down the stairs and swam half a kilometre.

My second pregnancy was completely different. As a working mum with a toddler at home, my life was more stressful, and my body had not yet fully recovered from my first pregnancy. This second time, everything hurt, I could not walk, I could not sit, and I could not lie down. I have deleted all photos of myself from that period, so I assume I looked as good as I felt. On top of that, the thought of giving birth again terrified me. My first childbirth had been a nightmare, so I had this cloud of anxiety floating above me. As if that were not enough, my husband was freaking out as he was so worried about how we would cope with two kids instead of only one. A well-founded, if not very helpful, worry.

This book is not about babies but about women. Therefore, we will focus on what happens to the woman's body during a healthy pregnancy, only getting marginal glimpses of the foetus' journey.

Pregnancy – the wonder begins

A pregnancy lasts around 40 weeks, and one starts to count the pregnancy from the first day of the last menses, that is before you are actually pregnant. You can count the pregnancy in weeks, months or in trimesters. People who have never experienced a pregnancy first-hand or as a related party, tend to count it in months but once you are pregnant, weeks are what you will focus on. The information you will get on pregnancy from different sources is most of the time presented in weeks.

The short version of a pregnancy is that the foetus implants and develops a placenta as a point of contact between itself and the mother. The foetus is fed through the placenta by the nutrition it extracts from the mother's blood. It also influences the mother's body by releasing hormones throughout the 40 weeks of gestation. Throughout the pregnancy the different pregnancy hormones – progesterone (pro-gestation) and oestrogen – skyrocket; also, other hormones such as lactin, relaxin and cortisol increase. All these hormones influence both the body, the mental state of the mother-to-be, and more or less controls her life during this period. It is also by controlling these hormones that the foetus decides when it is time to leave the comfy environment of the womb and venture out into the real world as a new-born baby.

But let us start from the beginning.

A lot of things need to be aligned to even become pregnant in the first place. Although the beginning of a pregnancy is counted as the first day of your last menses, you are not pregnant until the fertilized egg implants in the endometrium, the lining of the uterus. This happens approximately 8 to 10 days after ovulation, which is about 3 weeks into the calculated pregnancy.[448] So when a woman is 5 weeks pregnant, in the strict definition of the term she has only been pregnant for 2 weeks.

The uterus has checked the selection of the sperm and decided whether the early embryo was allowed to implant or not. However, once implanted, the foetus moves to the driver's seat and takes over the control of the pregnancy. A healthy foetus influences its own growth and development by sending messages to the mother's body, mainly through hormones. These hormonal messages impact the metabolism, the blood flow, the cellular changes and even communicate the foetus' desire and readiness to leave the uterus.[449] By doing so, the foetus does not only influence its own development but also has a direct effect the mother's body, driving most of the changes that happen to her. It may sound as if the foetus knows what it is doing, which of course is not the case. The whole process is already programmed into its genes.

Ever since the ovulation, the uterus has been preparing to host the embryo. Under the influence of the hormones secreted by the corpus luteum, the uterine lining has grown thicker, built up a network of blood vessels and created glands that secretes nutrients, a uterine milk which will nourish the embryo during the first couple of months of the pregnancy. This process – called decidualization – happens every menstrual cycle, pregnant or not.[450] The uterine lining plays an essential role in preventing the mother's immune system from expelling the implanted embryo from the uterus.[451]

Clinging-box

Once implanted in the womb, the first thing the foetus needs to do is to make sure it can stay by preventing the uterine lining from shedding through the menstrual bleeding. In order to do this when the foetus is only a couple of cells large, it starts secreting human chorionic gonadotropin (hCG).[452] The hCG prevents the corpus luteum to degrade as it usually does.

If the egg is not fertilized, the corpus luteum stops secreting progesterone and decays after 11–12 days. The first month of pregnancy, the corpus luteum grows to about twice its initial size and is just below 2 cm in diameter. About the size of a wedding ring. By secreting larger quantities of progesterone and oestrogen, the corpus luteum ensures that the endometrium continues to grow, and the survival of the foetus is assured.

The hCG is what is detected in the urine when you do a home pregnancy test. Within normal pregnancies, levels of hCG can vary widely from one woman to the next. Serum and urine concentrations of hCG rise exponentially in the first three months of pregnancy, doubling about every 24 hours during the first 8 weeks. The peak is normally reached around the tenth week of gestation, the levels then decreasing until about the 16th week of gestation, when they will remain fairly constant until delivery.

While celebrating Christmas with our family, my sister-in-law was pregnant without knowing it. With all the mouth-watering food also came a significant amount of good wine but, for some reason, my sister-in-law did not feel like drinking anything. Instead, it made her feel nauseous. The primary suspect of what causes such nausea in the beginning of the pregnancy is the levels of hCG.[453] The nausea – which is sometimes referred to as 'morning sickness' – probably has an evolutionary purpose so that women avoid eating things that can cause harm to the foetus, especially spoiled meat, but also alcohol and coffee for example.[454]

The placenta – not your organ

In the beginning of the pregnancy, the uterine lining shields the mothers main blood supply from the embryo. Hence, the growing embryo cannot access as much nutrients as it needs to sustain its rapid growth. To set up a direct connection with the mother's blood, it starts digging into the uterine lining. It is from the embryo's cell, and not the mother's, that the placenta is formed during the first weeks of pregnancy. The placenta is an organ of the foetus, and not of the mother.

The placenta acts as a central that allows an exchange to happen between the mother's blood and the foetus blood. It is a temporary organ attached to the uterine lining on one side and to the umbilical cord on the other. It is through the umbilical cord – which is formed after about 5 weeks of pregnancy – that the nutrition and oxygen will be passed over from the mother's body to the foetus. The umbilical cord has three blood vessels, one that brings the oxygen and nutrient rich blood to the foetus and two that returns the oxygen poor blood.[455]

After the placenta is fully developed – at about 12 weeks – it will gradually take over the hormonal production from the corpus luteum and generate large amounts of the two sex hormones, progesterone and oestrogen, produced to maintain the pregnancy.[456] The corpus luteum regresses slowly after 13 to 17 weeks into the pregnancy.[457] It is all these hormones that will drive the changes of the mother's body during the pregnancy but producing hormones is not the only task of the placenta. It also provides oxygen, nourishment and handles the waste. It is single-handedly doing the job of the lungs, liver, kidneys, and other organs until the foetus' own organs become functional.[458]

The placenta is usually attached to the top, side, front or back of the uterus. In rare cases – called low-lying placenta or placenta previa[459] – the placenta might attach in the lower area of the uterus as well, block the exit and create problems during the delivery.

Yummy-box

The Hollywood star Tom Cruise vowed that he would eat the placenta of the daughter he had with Katie Holmes in 2006. (A claim he downplayed in later interviews.) Two of the Kardashian sisters, Kim and Kourtney, made pills out of their placenta, pills they kept eating for a long time after birth. Even though some animals do it as well, there is no indication that eating the placenta of your offspring might provide any benefit for humans.

Some people think that eating the placenta can prevent postpartum depression, reduce postpartum bleeding, improve mood, energy and milk supply, and provide important micronutrients, such as iron, but scientists have been unable to furnish any evidence so far. Some well renown institutions, such as the Mayo Clinic, even publish warnings on such practices, because the placenta could contain infectious bacteria (they also specify that cooking it or preparing it in smoothies will not destroy the bacteria and viruses).[460]

Impact of the hormones – the communication channel

In the vampire saga Twilight, the baby Renesmee is able to communicate with her father while still a foetus in the womb of her mother (and hence telepathically inform them about her special diet). In real life, human foetuses have a direct mean of communication with their mother, less Hollywoodian but equally impressive: through hormones. Indeed, by sending hormones into the

Hormones during pregnancy

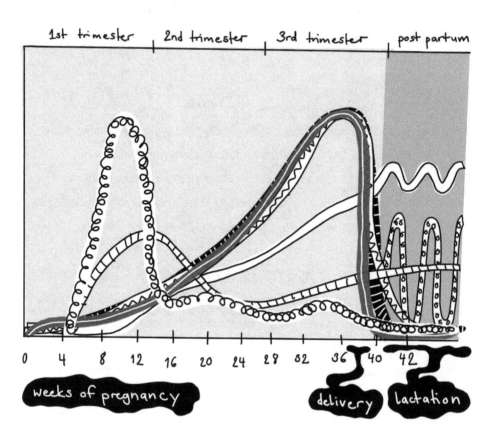

1st trimester | 2nd trimester | 3rd trimester | post partum

0   4   8   12   16   20   24   28   32   36   40   42

weeks of pregnancy

delivery   lactation

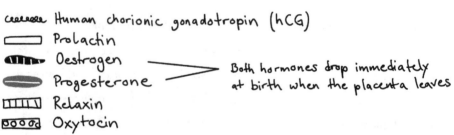

ceeeeee Human chorionic gonadotropin (hCG)

☐ Prolactin

◁▥▥▶ Oestrogen

⬭ Progesterone } Both hormones drop immediately at birth when the placenta leaves

⊞⊞⊞ Relaxin

ᓂᓂᓂᓂ Oxytocin

ᔨᔨᔨ Cortisol

mother's bloodstream, the foetus can get her to produce what it needs to grow and develop. The principal hormones used are progesterone and oestrogen that increase continuously over the pregnancy until a few weeks before delivery and then drops suddenly just after delivery.[461]

The high level of hormones produced and released by the placenta has a strong impact on the mothers' body. As we have seen in previous chapters, oestrogen stimulates cellular growth and during pregnancy, the extreme quantities of oestrogens make the uterus grow, as well as the breasts. The latter grow to almost double size and their ductal structure develops, making it possible to breastfeed once the baby is out. As the breasts increase in size, they can become sore and the area around the nipples becomes darker. Some researchers speculate that this could be to help the new-born babies, who have blurry vision, to easier find their way to the food. The oestrogen will also enlarge the genitals and relax the pelvic ligaments to make it easier for the foetus to travel through the birth canal.[462]

The extreme levels of oestrogen can sometimes have unintended consequences. A colleague told me about a friend, who started producing so much oestrogen during her pregnancy that it started to 'contaminate' her partner who also began to have high oestrogen levels. Can such a thing really happen and how would it work? It is not uncommon that men get hormonal fluctuations during their partner's pregnancy. This is called Couvade syndrome or 'sympathy pregnancy'. It was previously believed to be mainly psychological, but researchers have managed to show that the man's hormonal levels fluctuate in tune with the woman's: as their testosterone levels drop, the oestrogen and prolactin rise. Hence, men can get morning sickness, weight gain and mood swings.[463] This is probably what happened to the couple in question, but researchers are not entirely sure about the mechanism behind this phenomenon.

A strange hormonal effect happened to another friend. She has been pregnant three times, and she admitted shamefully that every time she hated the smell of her husband. It is a common side effect of the fluctuating oestrogen levels that the taste buds and the sense of smell change, thus modifying which food you like and sometimes giving you a strange taste in your mouth. A metallic taste is the most common, but also a salty, burnt or foul taste has been reported.[464] This is called dysgeusia and the resemblance to 'disgusted' is not a coincidence. It is the same phenomenon that is behind the strange pregnancy cravings you often hear about. Dysgeusia is normally at its worse during the first trimester of the pregnancy.

Before implantation of the foetus, progesterone plays a very important role as it changes the environment in both the fallopian tubes and the uterus to facilitate the implantation. Once pregnant, it keeps playing a key role as it relaxes smooth muscles. This is particularly important since it prevents contractions of the uterus from expelling the foetus. Just as during the luteal phase, this relaxation effect stretches to the intestines. It slows down the digestive system, making the absorption of nutrients more efficient. On the downside, you become constipated, gassy, bloated and get heartburn.[465] The secreted progesterone has another major role to play – during the pregnancy it also teams up with oestrogen to prepare the breasts for breastfeeding.

In the beginning of the pregnancy, it takes some time for the body to get used to the high progesterone levels, which is why it is common to be very tired in this phase.

Energy-box

During pregnancy, the placenta produces and releases a special hormone, the human placental lactogen. One of its important functions is to change your metabolism so that the foetus can receive more energy. It also makes your body less sensitive to the effects of insulin, which leaves more glucose available in your bloodstream to nourish the foetus.[466] This can in some cases lead to gestational diabetes,[467] which is diabetes that is diagnosed for the first time during pregnancy. In most cases the diabetes will go away shortly after delivery but when it does not, it will be classified as type 2 diabetes.[468]

Shake-up-box

Almost all endocrine glands react to pregnancy in some way, either due to the increased metabolic load or as a consequence of the hormones released by the placenta. The anterior pituitary gland – which is normally a part of the hormonal axis formed together with the hypothalamus and the ovaries – reacts to the increased amount of sex hormones from the placenta by completely shutting down the production of follicle stimulating hormone and luteinizing hormone, which will halt the menstrual cycle. On the other hand, the pregnancy increases the production of many other hormones, among others prolactin, which will become important to stimulate breastfeeding.[469]

My feet grew about half a shoe size per pregnancy, and it was only my husband's refusal to have a third child that saved me from buying clown shoes. This was not only due to the constant weight that you carry around, but also to the hormone relaxin, that is released during the pregnancy. Relaxin is a hormone that can be produced both in the ovary and in the placenta. During the menstrual cycle, it is produced by the corpus luteum and increases during the second part of the menstrual cycle, where it is believed to play a role in preparing the uterus for implantation. During pregnancy, the role of relaxin is, like the name implies, to relax the ligaments and the bones in the pelvis to help giving birth. In many women this leads to pain and discomfort in the pelvis and lower back during the first trimester,[470] but the effect is not completely local. Hence, the feet are affected and get looser and more spread out. These effects are not reversible, and I simply had to buy new shoes.

CRH-box

In this hormonal turmoil, another player appears which increases during the whole pregnancy: – the corticotropin-releasing hormone (CRH) – also released by the placenta. This will in turn lead to increased levels of the stress hormone cortisol. Cortisol levels in the third trimester of the pregnancy are about three times as high as they are in non-pregnant women. In spite of the very high levels of stress hormone, women in late pregnancy are not acting particularly stressed, thanks to the downregulation of the hormonal axis responsible for the stress response.[471] At birth, the CRH levels drop immediately with the delivery of the placenta, but it will take another 3 weeks or so until the cortisol return to its normal levels.

The CRH will play an important role, both in the development of the foetus and in the behaviour of the mother.[472] It is instrumental in regulating the blood flow between the placenta and the foetus,[473] it helps the foetal organs mature, and also appears to influence the timing of birth. Towards the end of the pregnancy, the increasing cortisol levels seem to play a role in the brain development of the foetus as well as the maturation of its lungs.[476, 477]

After the birth, CRH still has a busy schedule, with its important role in how mothers bond with their babies. Researchers have shown that baboons mothers with higher levels of CRH during pregnancy spend more time watching, grooming and manipulating their babies.[478] Similar results have also been found in humans, although the exact mechanisms are still to be clarified. So far, scientists have surmised that the hormones could have a direct effect on the brain, making us more vigilant and emotionally aroused,[479] or that the presence of CRH is simply a marker of other hormonal changes that promotes bonding.[480, 481]

The thyroid and the parathyroid gland become much more active during pregnancy and breastfeeding. The parathyroid gland ensures that the foetus, and later the breastfeeding baby, gets enough calcium to grow its bones. If the mother has a diet poor in calcium, the parathyroid will 'steal' calcium from the mother's bones to make sure there is enough in the fluid that passes to the foetus through the placenta, or later to the baby through the breastmilk. Since bones are a living tissue that regenerates every 3 months, it is possible to build up the bone again with a diet rich in calcium and vitamin D.

Cardiovascular and respiratory system

It is of course not possible to have a little someone hacking into your system and living from your blood without sustaining major changes to your cardiovascular system. Hence, during pregnancy, the blood volume increases significantly, and the blood vessels dilate, mainly due to the amount of progesterone.[483] Even the heart's position changes, with a slight rotation.[484] The pulse rate increases continuously over the whole pregnancy and is on average almost eight beats per minute higher towards the end of the pregnancy than at its beginning.[485]

For most women blood pressure decreases over the pregnancy.[486] In some it can however increase. Within a certain span it is still considered normal but could be a sign that something is wrong if outside this range, like preeclampsia, which is a very serious condition. As many as 35% of all women with hypertension in pregnancy develop preeclampsia.[487]

Preeclampsia-box

Preeclampsia is believed to be initiated by abnormalities in the placenta, causing damages to the vasculature. This would explain why in most cases the symptoms disappear immediately after giving birth. This vascular detriment in return harms several other organs such as the kidneys, the liver, lungs, blood vessels and more.[488] The damage in the kidneys impairs the filtering function and proteins in the urine. If preeclampsia is not immediately treated, it can turn into eclampsia, which is the onset of seizures.[489] Eclampsia derives from the Greek word meaning 'sudden development' and refers to the fact that it appears so suddenly. Preeclampsia can be seen as the high-risk strategy of the foetus to get more nutrients. The mother's blood vessels become constricted thereby causing high blood pressure and more blood with nutrients to be sent to the foetus.[490] As the placenta is in fact an organ of the foetus and a genetic mix of both parents, it is maybe not surprising that the father might play a role in the development of preeclampsia.[491] Nevertheless, the role of the father is controversial. It seems that there is a correlation between risk of preeclampsia and changing father between the first and the second child – if the mother did not have preeclampsia with her first child, the risk of preeclampsia is higher with the second child if it is with a different father. However, as the inverse correlation is not true – if the mother has preeclampsia with her first child, the risk is not lower if the father is different for the second. Probably a long interval between two deliveries is more important than a change of paternity. This casts a shadow over this 'dangerous father' theory.[492]

One of the marvels of modern medicine is how the risk of being pregnant and giving birth has been reduced from being a common cause of death for women, to something extremely rare. During the nineteenth century 500–1000 women died of childbirth for every 100,000 pregnancies. Today those numbers are drastically lower. In Europe the risk of dying while giving birth is around 4–8 per 100,000 pregnancies but depends slightly on the country.[493] This used to be the same in the USA, but according to the CDC the mortality rates have gone up from 7 in 100,000 (within Europe's range) in 1987, to 17 in 2016.[494] There is a significant difference between ethnicities, with over 40 black women dying per 100,000 versus 11 for Hispanic women and 13 for white women. The major risk factors are pre-existing conditions and especially cardiovascular complications. Despite being very rare, they are the leading cause of maternal deaths nowadays. In the USA, such complications cause a fourth of all deaths during pregnancy or postpartum. In particular hypertensive disorders – an umbrella term also including preeclampsia and eclampsia – affect as many as 1 in 10 women and cause increased risk of heart failure, heart attack and stroke. African-American women are especially exposed with risks almost three times higher than Caucasian women. Being pregnant over 40 is another strong risk factor.[495]

Weight gain – competing with polar bears

Polar bears are most likely the champions of weight gain in pregnancy. Starting from an already impressive weight of 150–250 kg, they almost double their weight and must put on at least 200 kg for a successful pregnancy.[496] This weight is crucial to be able to properly nurse their cubs during the long hibernation.

Even though I cannot compete with a momma polar bear, I shone in my own league, putting on quite a lot of weight during both my pregnancies. For the first one, I managed to stop exactly at the upper limit of what was recommended for my weight category but surpassed it for the second. The recommended limit is of 11 to 16 kg, with a BMI ranging between 18 and 25. If you are lighter, you should put on a bit more weight, and if you are above that span, you should put on less.[497] I gained 16 kg during my first pregnancy and, while pregnant, I felt great, not feeling fat at any moment. However, when my daughter arrived and I still had 12 of the 16 pregnancy kgs left, it was not at all as comfortable.

During the delivery, I got rid of 4 kg. Of those, the baby was about 3 kg (3.1 kg in the case of my daughter), the placenta is about a sixth of the weight of the baby – so let us say 0.6 kg[498] – and the rest was amniotic fluid. The average amount of amniotic fluid at delivery in week 40 is 800 ml, which would weigh 0.8 kg.[499] All of this added up to a total of 4.4 kg. So where did all the other kilos come from?

The uterus – which generally put on 1 kg or so – accounts for some of the additional weight; the breasts increase in the same proportion. About 4 kg is fluid that has accumulated in the different tissues and in the increased blood volume;[500] this you will pee out during the days following birth. What is left is the weight that is so difficult to get rid of, but those extra kilograms in energy storage might become handy when you need a lot of additional calories for breastfeeding. The first couple of weeks after my daughter's birth, my husband was at home with me and then my parents picked up the torch. But then everybody left, and I was all alone with my daughter: the problems started. I had to take care of the baby – a job I was learning by doing – and of myself. Since she was the priority, I ended up starving. I did not manage to cook or even make myself a sandwich during the day, which meant that I did not eat until my husband came home from work. I guess that – just like the momma polar bears – that additional weight helped me produce sufficient amounts of nutritive milk for my daughter despite the circumstances.

Weight gain or not, it is necessary to eat more during pregnancy. The metabolic rate of a pregnant woman increases by about 15% during the latter half of pregnancy. Therefore, you might often get the sensations of becoming overheated in pregnancy. Also, by simply carrying around a lot of additional weight, your muscles will use more energy than they normally do. We also get hungrier during pregnancy since food substrates are removed from the blood to give it to the foetus.[501]

During the last months of pregnancy, the nutritional needs of the foetus are skyrocketing (remember, we are talking about creating life from scratch). The mother's body cannot get enough protein, calcium, phosphates and iron; luckily, it has anticipated this issue and has been storing these substances, either in the placenta, or in the normal storage depots. However, if the diet does not contain enough of the appropriate nutritional elements, a number of deficiencies can occur, especially in calcium, iron and vitamins.[502] Most of the calcium is stored in our bones and if there is an overall lack of calcium, this is where the body will extract what it needs. A calcium deficiency can therefore have detrimental effects on bone health.[503] Iron deficiency can make you weak and tired if it is left untreated. It can cause anaemia which can result in premature delivery and low birth weight for the baby.[504] Each vitamin has an important role in supporting the growth of the foetus.[505]

If having sufficient fat reserves is necessary for a healthy pregnancy, too much weight gain can be harmful. This increases risk for disease such as hypertension, gestational diabetes, complications during labour and delivery, as well as unsuccessful breastfeeding.[505]

The moving organs

Have you ever seen one of those online videos showing how the organs get pushed around by the growing uterus during pregnancy? I would recommend the experience; it is rather mind-blowing. The growing belly that you can see on the outside is just one part of the story; on the inside there is a complete re-ordering of the organs to make room for the foetus and the placenta (remember, we are talking about 4 kgs towards the end of the journey). The small intestine gets more and more compressed and pushed upwards, whereas the bladder is squeezed towards the bottom of the pelvis, which can lead to incontinence (or, at the very least, frequent trips to the toilets). Towards the end of the pregnancy, the abdomen is definitely overcrowded, with most organs being compressed, stomach and lungs included. Many of the pregnancy symptoms are due to the changing hormones, but squashed organs have their fair share of responsibility. The lungs are impacted by pregnancy too, both due to the increased need of oxygen for the foetus and to the mechanical influence of the growing belly pushing the lungs upward. The respiratory rate does not necessarily change but the amount of air passing through the lungs augments.[506]

I was recommended to lie down as much as possible for the two weeks following the birth of my children or, I was told (or think I was told, my German was not so good at the time), my organs could 'fall out' of my body. This, somehow, made sense to me considering the extreme reordering the organs had been subjected to. And since I was rather worn down, being ordered two weeks bed rest was exactly what I wanted anyway. Bed rest is not recommended everywhere since it can increase the risk of pulmonary embolism and thrombosis but taking it easy and being relaxed is.

The pelvic floor is a group of muscles spanning from the pubic bone in the front to the tail bone in the back supporting the inner organs – uterus, vagina, bladder and bowels. It is put to considerable strain by the burden of the growing foetus and placenta, and then by the extreme stretching endured during birth. Therefore, some issues after giving birth are normal. The pelvic floor has been weakened but, since it is a muscle, it can be trained again. It takes from a couple of weeks to several months before it will back in shape.[507] That being said, many women struggle with their pelvic floor for years after giving birth.

Pelvic organ prolapse occurs when the organs 'fall down' or slip out of place inside the pelvis. Pregnancy and vaginal birth are the main risk factors, but not the only ones: aging, menopause, smoking and chronic diseases can also lead to it. After the delivery, prolapse is a consequence of the relaxation of the ligaments holding the organs.[508] Childbirth is the biggest, but not the only contributing factor to pelvic prolapse. Risk factors for the development of pelvic organ prolapse include aging, menopause, smoking and chronic diseases. This risk of prolapse is proportional to the size of baby, and especially critical beyond 3.8 kg.[509]

The strain of pregnancy itself is not the only thing that can harm the organs. Vaginal delivery can also lead to a high rate of incontinence in the postpartum period. Women who had no issues with bladder control during the pregnancy can find themselves with stress incontinence after the birth. Stress incontinence is when you leak urine during certain movements or activities, such as coughing, laughing, running or heavy lifting. In fact, the reported incidence incontinence after a first vaginal birth is of 21% with spontaneous birth, and of 36% when the birth had been assisted with additional instruments (forceps). Stool incontinence happens to about 2–6% of the women. Sexual function can be a problem immediately after the birth, but it will usually recover.[510] The incontinence does not always go away if left untreated, so it is important to start training your pelvic floor as soon as you can after delivery.

Amniotic fluid – the water that breaks

The pregnant mother's water breaking is often a highlight – of emotions and chaos – in movies (you can even find lists of the best water breaks on the net!). It is dramatic, splashy, and usually happens at a most inconvenient time. My water breaking was much less theatrical: it occurred when I was already hours in the delivery and, for the second birth, only minutes before my son came out.

The 'water' is in fact the amniotic fluid that is surrounding the baby inside the uterus and that forms a protective cushion around it. In the beginning of the pregnancy, it mainly consists of water, salts and minerals made from both the mother and the foetus. Late in the pregnancy, the urine of the foetus will become the main component of the fluid.[511] Amniotic fluid contains live foetal cells and other substances, such as a special protein produced by the foetus (alpha-fetoprotein (AFP)). These substances provide important information about your baby's health before birth. It is the reason why a sample of the amniotic fluid can be used to test for certain birth defects such as Down's syndrome.[512]

At the time of giving birth, a friend had 5 litres (allegedly, but let's say a lot) of amniotic fluid, which is about five times the normal amount. This complicated the birth greatly, and a whole squad of hospital staff had to assist. When her water broke, it was like emptying a whole bucket over the entire personnel, and one of her strongest memories was the sound of the doctors trampling around in wet shoes, splashing water around as they were helping her. This excessive accumulation of amniotic fluid is called polyhydramnios and occurs once or twice in a hundred pregnancies. There are a few known reasons for polyhydramnios but most of the time, the reasons behind it are not clear. The earlier that polyhydramnios occurs in pregnancy and the greater the amount of excess amniotic fluid, the higher the risk of complications. Luckily, my friend got away with a scare and the guilt of wetting her doctor's shoes.

Brain and mood during pregnancy – not all harmony

Another friend of mine, while pregnant, would completely freak out in anger towards her husband, lock the bedroom door and peacefully fall asleep on her own. The morning after, she would wake up in a great mood and hardly understand what had happened and why her husband was sleeping on the couch.

Such mood swings are not uncommon in pregnancy. A pregnancy can be quite an emotional rollercoaster, dashing from tears one moment to euphoria the next. Some of the mood swings can be attributed to the new situation and the anxiety brought about becoming a parent. Just as the hormones affect our mood in the menstrual cycle, they also influence us in pregnancy. Your reactions during your menstrual cycle could provide some indication to how your mood will be impacted during pregnancy. In the case of my friend, progesterone was most likely taking its toll. The adjustment to the new hormone levels can be especially tiring in the beginning of the pregnancy.

Maybe because I am a person who gets very happy during my high oestrogen phases in my menstrual cycle and does not have too much negative influence of progesterone, the pregnancy was a pleasant time for me, but I had problems with anxiety. I was continually worried that something would go wrong with my baby. This constant state of tension is commonly widespread among mothers-to-be, and many of my friends were prohibited to go on online forums by their doctors, as that would make them even more anxious. Some level of anxiety is completely normal and not necessarily linked to the hormonal changes. However, there are structural changes that happen to the brain during pregnancy that influences the level of anxiety. Through studies of how different brain regions are activated in new mothers, researchers have found that the regions for nurturing and detecting risk were further evolved and more activated.[513]

The concept of a 'baby brain' is one of the myths surrounding pregnancy. A 'baby brain' means the alleged decrease of cognitive skills during pregnancy and the early days of motherhood. This includes memory problems, poor concentration and absentmindedness. There is no consensus on this topic, with studies claiming such changes occur in a home setting but not in a professional one, others asserting nothing of the sort happened at all. There is therefore no reason to assume that you would experience a cognitive decline in this period.[514]

It could be that young mothers and mothers-to-be simply have different perspectives on life and choose – consciously or not – to focus on topics that are important to them at that moment. Things that mattered to you before being pregnant can somehow feel very trivial and uninteresting, when you grow a life inside of you and wonder about the huge life changes that are ahead of you.

Nevertheless, the brain might be influenced by pregnancy since cells of the foetus can actually be found in the mother's brain.[515] How those cells exactly impact the mother's brain is yet unknown, but some researchers have posited that it could be something that makes us recognize our offspring better.

Composite-box

A chimera is a creature made up of cells of two or more different individuals. Sounds like science fiction or even mythology? It is not. It turns out that mothers may have kept some of their children's cells in their body. It is cells that have migrated from the placenta into the mother's body, and they can be found everywhere. In women that bore sons, you can for instance find cells with Y chromosomes in their brain, lungs, spleens, livers, kidneys and heart.[516] As we are only talking about a small number of cells in this case, it is called microchimerism. The purpose behind this is not yet known, but when you say that mothers always carry a part of their children with them, it is literally true.

Giving birth – a subjective experience

When my daughter was about five years old, she asked me if it hurts to have a hole in the leg. Very puzzled, I asked her what kind of hole she meant, and she said: 'You know the kind of hole babies come from.'

'Aha', I answered, 'but they come from a hole between the legs, and not in the leg.'

'Oh, but does it hurt?'

I said 'Yeah, it hurts pretty much...' (and I thought 'Oh my, I didn't know one could survive such pain!').

Then she burst into tears and cried: 'I never, ever, want to have any babies!'

I get her point. If it were not for the long-lasting love story, you will build with your children, no one would go through such an ordeal.

Whereas women apprehend their pregnancy in many different ways, their experience of giving birth spans across an even wider range, from heaven to hell, but in my opinion, surprisingly many talk about it as a beautiful moment. Indeed, if a majority of women experience an awful lot of pain during childbirth, 3 months later up to 90% of them look back on it as a satisfactory event.[517]

I have a hard time finding labour itself a positive experience. My best comparison would be being tortured under the Spanish Inquisition. That being said, I also met the love of my life by giving birth to her, so it was absolutely worth every minute of pain I had to go through. What was it that I found so traumatising? Medically, I had a very successful birth. But the pain that was so overwhelmingly atrocious. I had no idea one could feel that (and survive). When the contractions started, I was at home and every 10 minutes I would have a painful contraction but totally withing the range I could deal with. I did everything by the book: I moved, I breathed, and had a feeling of control. This continued for about 7 hours with, as expected, shorter and shorter intervals between the contractions. And as they should, the contractions got more painful and lasted longer and longer. By this time, we were in the hospital. Up to that moment, things had been going decently. I still felt in control and I kept on breathing and moving as I should. But it was getting worse and worse and soon the time between the contractions was so short that I had no chance to recover in between. The pain was excruciating and the breaks non-existent. It was like my hips were tied to a medieval stretch bench that was, slowly but mercilessly, pulling my pelvic bones apart. At that point, in a haze of ache, I thought that I would rather die than take another minute of this.

Eventually, I managed to convince the midwife that I needed an epidural and when the anaesthesia finally kicked in, I immediately came out of my cloud and it felt amazing. I looked at my husband, whose existence I had completely forgotten about; he was pale as a ghost. He was in complete shock of having seen me suffer like that and my epidural needed a longer time to influence his mental state than my body.

From this moment on, the birth became a tolerable experience, even beautiful. This was 11 hours after the first contractions. By then, it was evening, and the lights were low in the delivery room. We could bring our own music and I had brought my favourite albums of Morcheeba that were now playing in the background. The pain was still present but manageable, and I could be mentally present and understand what was happening. The midwife put a mirror so that I could see the head of my daughter coming out. That was a truly wondrous moment and the beginning of a new life for us as a family.

I am not telling you this story to scare you off giving birth. I am telling you this because I believe that a part of my trauma was that I had a preconception that a strong, healthy, woman like myself would be able to go through labour without any pain relief. Before giving birth, I was always told positive stories about resilient women having natural births in harmony with their bodies. I considered the fact that I ended up using an epidural as a failure. The attitude of the midwife did not help, to say the least. But getting help is not a failure and giving birth with or without pain relief is not a matter of being courageous or not. I will later go through the different factors that will influence the amount of pain and how it can vary largely from one woman to the next, making comparison impossible. I also want to emphasize that, no matter how painful you will find giving birth, the reward will always be worth it. But before coming to the moment of giving birth, many steps need to be taken in the preparation of welcoming the new-born.

Preparing the body to give birth

Towards the end of pregnancy, the body starts preparing for childbirth, and the uterus starts practicing contractions. These 'false' labour pains that occurs before 'true' labour starts are known as Braxton Hicks contractions. It is periodic episodes of weak and slow rhythmical contractions. They may start already in the second trimester but are more common in the third trimester of pregnancy. Contrary to true labour contractions, the 'false' labour contractions will stop when you change position or relax. Toward the end of the pregnancy, the contractions become progressively stronger and, in the end, they change to become strong enough contractions that start stretching the cervix and force the baby through the birth canal. During pregnancy, progesterone inhibits uterine contractility to prevent too early expulsion of the foetus.[518]

As another step towards delivery, the cervix gradually softens. In non-pregnant women, the cervix is closed and firm, and its consistency is similar to cartilage. By the end of pregnancy, the cervix has become easily expandable and more compliant – with a consistency similar to the lips – but still firm and unyielding.[519] Beginning a few weeks before labour, the cervix begins to ripen, with a significant increase in blood vessels and growth of different cell types. A type of inflammation is involved in the ripening of the cervix, and if it is for some medical reason is necessary to induce labour, prostaglandins are provided to speed up the ripening of the cervix.[520]

Triggering childbirth – what really works

The day before giving birth to my firstborn, only a few days before the due date, I was still at work. I felt heavy like a hippo, and my mind was everywhere except on the algorithms I was working on. Instead, I was searching the internet for all possible ways to trigger the birth so I would not have to go to work the following week. I found a lot of information about what I could do and, as a true scientist, I decided to try them all. Abandoning my workstation, I pretended to go to the bathroom, and instead walked all the stairs I could find. Up and down, up, and down. Around lunchtime, I gave up and went to the pool, where I swam half a kilometre. I then went home and drank about a litre of red raspberry leaf tea and took a nap. After that, I went for a two hour walk with a friend in the forest and went home to cook a very spicy meal and yes, we also had sex. Basically, I followed the midwife handbook on how to induce labour, and it worked. The morning after, the contractions started.

Did it actually work? Since it was just two days before the estimated term, there is no way of knowing if my attempts worked, or if my daughter was simply ready to come out. Scientifically, only a few of those things make sense, whereas others are just 'old wives' tales'. There is for instance no evidence that spicy food should in any way induce labour. Although many recommend drinking red raspberry leaf tea, there have been no scientific findings to corroborate this advice.[521] The same goes with exercise, which is on the top of the 'let's induce labour' list.[522] Sex on the other hand could have helped, because it releases the same oxytocin that plays an important role during the birth itself. Oxytocin is a hormone that is administrated to medically induce labour. Its role in a natural birth is however not clear. The oxytocin will stimulate the uterine muscles to contract and increase the production of prostaglandins, which in turn increases the contractions even further.[523] The foetus' pituitary gland also secretes oxytocin which could potentially play a role in kicking off the labour.

Time-box

Exactly what determines the timing of the birth is not fully understood, but it is most likely a combination of signals from both the mother's body and the foetus. Just like prostaglandins play an important role in menstruation where they cause the uterus to contract, they also play an important role in the induction of the labour. It is the involvement of the prostaglandins, which among others cause inflammation, that make some researchers talk about both menstruation and labour as inflammatory events. During the whole pregnancy, the levels of progesterone have been continuously increasing. Progesterone acts as an anti-inflammatory substance by suppressing prostaglandins and oxytocin. It also keeps the uterus calm and relaxed. Towards the end of the pregnancy progesterone decreases and this leave room for the prostaglandins to act to trigger the inflammation that starts the labour.[524]

Others claim that the initiator is in fact a signal from the foetus itself. Researchers at UT Southwestern Medical Center have identified two proteins that are secreted by the foetus' lungs indicating that they are completely developed and ready to function in the outside world. The proteins are then secreted into the amniotic fluid, leading to an inflammatory response in the mother's uterus that initiates labour.[525]

The placenta also tells when it is time for this pregnancy to end.
This is done by the CRH that is released by the placenta and has increased over the course of the whole pregnancy and peaks at the very end.
If CRH is playing an active role in triggering the birth is not proven, but we know it is closely linked to cortisol, and that the cortisol produced by the foetus' adrenal gland plays a role in the inflammation of the uterus.[526]

Oestrogens do not cause contractions, but they seem to prepare the uterus and coordinate and enhance contractions.[527] In other species, oestrogen promotes the birth to start. In humans, it is unclear how sensitive the balance between progesterone and oestrogen is, since oestrogen levels also increase over the entire pregnancy, but it is possible that when progesterone drops and the balance between oestrogen and progesterone is dominated by oestrogen, the contractions can be triggered.[528]

Early-box

Early preterm labour can be caused by infection, as well as by microbes.[529] According to the Centers for Disease Control and Prevention (CDC), one in nine births in the United States in 2012 occurred before the 37th week of pregnancy. Complications following such premature births cause 35% of all infant deaths, making prematurity the leading cause of death in babies.[530]

There are more cases of early births now than in the 90s. Why is that so? One clue might be the level of the stress hormones CRH and cortisol. Women with high levels of CRH in early pregnancy tend to deliver earlier whereas those with particularly low levels tend to deliver later, and even past term. In the US, Caucasian women have about a 9% risk of premature birth in 2018, while African-American women's risk is 14%; the latter may be related to socio-economic factors.

Scientists have observed a significant drop in the number of premature births in 2020. So far, they do not know the exact reasons, but think that the pandemic of Covid-19 might bear some responsibility. They have ventured that staying at home, getting more sleep, and being surrounded by your family, might alleviate the general state of stress, and thus decrease the levels of CRH and cortisol. On the other hand, confining yourself at home to avoid corona also helps to avoid other infections which are known causes to premature labour. The extraordinary – and, in our technological age, unprecedented – situation prompted by the pandemic and its lockdowns is becoming something of a full-size experiment. By studying the trends in the data, researchers might be able to understand why premature birth happens in general and how to avoid it.[531]

Delivery, act one: the show begins – contractions and cervical dilatations
My own delivery started with a sudden onset of strong contractions in the morning when I woke up. It is a rather commonplace scenario, another one being the 'bloody show'. That probably sounds more dramatic than it is. The 'bloody

show' is the release of a mucus plug and some minor amounts of blood that had blocked the passage to the uterus during the pregnancy.

During the first stage of labour, the cervix opens. This opening is driven by the contractions which start with a 10-minute interval and end with less than a minute during the second stage of the birth. Each contraction lasts from between 30–90 seconds. The cervix needs to open to a minimum of 10 cm in diameter to allow passage for the baby. The cervix opens between 0.5 to 0.7 cm per hour, a pace that is very individual and can be increased when the route has already been followed by an elder sibling.[532]

If the muscular contractions are not painful, the uterine ones are a terrible ordeal. Scientists are at a loss to explain why – and by now really you must have gathered that there is still much to be understood and learned. One theory is that the nerve endings in the cervix and lower uterus are compressed.[533] The intensity of the pain – from "Oops, I stubbed my toe on the desk!" to "Ouch, someone just tore off my arm" – vary a lot from one woman to the next. The pain is related to the speed at which the cervix dilates and correlates well with intensity of the contractions, how long they last and how frequent they are.[534] According to my midwife, my cervix was opening very fast for someone giving birth for the first time, which could account for my dreadful experience. Data show that women giving birth during the day tend to rate the pains lower than women giving birth during the night.[535]

Simulator-box

Would you ever be tempted to know what it feels like to give birth there are pain labour simulators that you can test. And if you just want a good laugh,

YouTube is full of men doing just that. Although it is mostly for the fun of it, they all come out with a whole new respect for women that have gone through the real deal.

Laughing-box

You are not condemned to go through the pain if it becomes too much. There are several options. The natural methods are breathing, relaxing and moving.[536] During my labour, I was offered nothing between the natural and the epidural. An epidural is a type of local anaesthetic. It numbs the nerves that carry the pain impulses from the birth canal to the brain. The chest, tummy and legs may feel numb while the epidural medicines are used. For most women, an epidural gives complete pain relief.

Often there are several other options between no pain relief to the heavy artillery that I got. One that seem particularly nice, is laughing gas. Laughing gas, or nitrous oxide, has been used as an anaesthesia ever since the American dentist Horace Wells started using it for pain relief while pulling out teeth in 1844.[537] All my Swedish friends swore by this gas when they were giving birth, so I was very disappointed that this option was not available in the Swiss hospital where I gave birth.

Act two: the end of the tunnel – foetal descent and delivery

During the second stage of labour, when the foetus is on its way to the birth canal, the levels of pain intensify dramatically because it is stretching the tissues on its way out; the cervix, the perineum, and also the stretching or tearing of the vaginal canal itself.[538] The stretching of the cervix further increases the contractions of the uterus.

Contracting-box

The more the baby descends in the birth canal, the more the cervix stretches and the stronger the contractions get. The exact cause is not known.

It could be a positive feedback system, or a mechanism where the stretching of the cervix stimulates the production of oxytocin from the pituitary gland, which also accelerates contractions.[539]

If the contractions halt at some point – which happened to me when I finally got the epidural – the doctor or midwife sometimes decides to rupture the membranes so that the water can flow out. This speed up the contractions again, probably because the baby falls deeper in the cervix and thereby stretches it.[540]

The uterus is made of two parts, an anatomical fact that becomes manifest during birth. During the contractions, the upper part becomes hard while the lower one – together with the cervix – becomes soft and dilates during the contractions. The upper part does not return to its initial shape but stays shorter as it was during the contraction. When they scream 'Push, puuuuuuush!' in movies, what you actually have to do is to use the muscles in your abdomen and give that upper part of the uterus an additional push to get the baby out. In most cases, you will have a natural urge to push at that moment anyway.

No birth is like any other, not even for the same woman. The first time I gave birth, it was a long and tedious process. The second time, I came into the clinic on all fours and, traumatized from my first birth screaming for an epidural, while my husband was parking our car. The staff asked if I preferred to wait for the anaesthetist or to give birth immediately and, before I knew it, my son was born. For her second baby, a friend of mine rushed into hospital, and hardly had the time to pull down her pants before giving birth with her big winter boots still on. My former boss had to help his wife deliver three of his four kids in the car, on the way to the hospital. Such stories are quite common when it comes to the second and third child. The time difference is due to the dilatation of the cervix. For a mother giving birth for the first time, the baby's way through the cervix is slow and steady. The following time this can go much faster since the cervix has not fully recovered its initial state.[541]

Now, the baby is out, but the mother and the baby are still physically connected: what happens next?

Cutting the strings – umbilical cord

As you know, – the umbilical cord is the link connecting the baby to the placenta. When the baby is out, the placenta is still inside the mother's womb and the two are still physically joined through the umbilical cord. To wait some time before clamping and cutting the link is beneficial for the baby. It allows for more blood to flow from the placenta into the baby.

A friend gave birth to her second child on her own in the parking garage of the hospital, while her husband had gone to find her a wheelchair. Despite the challenging context, she remembered that you should keep the baby low, at the level of the placenta, in order to have the blood running in the right direction. It was not a terribly comfortable position – with one foot still inside the car, one on the concrete floor and the baby in the trousers – but luckily someone walked by and held the baby for her while waiting for a more professional help.

Whilst we can marvel at her presence of mind, the baby would have been fine even if she had not managed to think about such details. For long, specialists recommended to hold the baby at the level of the vagina for at least a minute. It was believed that gravity somehow helps the blood flow in the right direction, but recent studies have shown that this is not necessary.[542] Although slightly terrifying, the story about my friend had a happy ending, and everyone is safe and sound. The baby is no longer a baby but a young and healthy boy, running around playing in the yard.

Act three: placenta leaves the scene

After my daughter was out, she first spent a few moments on my chest, and then she met her impatient and delighted father for the very first time. As she was safely resting in his arms, it was time for the last stage of the birth: delivering the placenta. A doctor was present together with an apprentice, the former explaining me and the latter (the father was not paying attention at this time, all smitten with his daughter) what was going to happen. She was going to massage my belly in a specific way and the placenta would come out. I must admit that I did not feel much at this stage, but she must have done her manoeuvre very successfully. While I was lying there, looking between my legs at those two women who were staring into the massacre of what was left of my nether parts, the placenta squirted out, and suddenly the faces of the doctor and her apprentice were splattered with blood. Luckily, they both had their mouths closed.

Bloodbath-box

Before the placenta comes out, it needs to be detached from the wall of the uterus. This is the most hazardous stage of the delivery for the mother. Since the uterine wall and placenta are so closely interconnected there is a risk for severe haemorrhaging of the uterus which can lead to a dangerous blood loss. Sometimes, the placenta grows too deep into the uterine wall – a high-risk condition called 'placenta accreta' – which can cause severe blood loss after delivery. Fortunately, doctors are generally able to detect it during routine ultrasound and often choose to perform a caesarean section.[543] Healthy women can manage a blood loss of up to 1000 ml. In low-income countries though, women often being already anaemic, they cannot afford to lose more than half a litre.[544]

Life after the partum

Emotional overload and bonding

I was a bit surprised about the feelings I got when I first met my new-born. It was not love at first sight as some parents advertise. After all, I had never seen this creature before and carrying someone in your womb is a very abstract thing. During the pregnancy, it was difficult to imagine that there was a real person inside me, and when I finally got to meet her, I felt mostly fascination and a strong curiosity about who she was. I wanted to get to know her, and I had a strong urge to take care of her, but it was not love. During the first 24 hours, I had this crazy, obsessive feeling that I wanted to have her all to myself and not share her with anybody, ever. Running away from my husband, family and friends and setting up a new home in full isolation crossed my mind several times. I felt like one of the chimpanzee mothers that leaves her troop to give birth and fully concentrate on her new-born baby without any interference of others. Who knows, maybe it is a leftover from our ancestors that we, just like the chimpanzees, feel the drive to leave to protect our offspring from being eaten by other troop members.[545] Eventually, I stuck around and, before long, I was super grateful for the help from others. It took a few days, and then the feeling of love also came about. Her father – who shared this need to protect her at all costs – was developing his own relationship with her with just as strong feelings as I had.

Of course, not all new mothers feel the same. There are probably as many different reactions as there are women. A friend of mine had no problem going to the restaurant with her husband and leaving her infant in the hospital just one day after giving birth, and she was always just as close to her child as I was to mine. We all react differently.

These new obsessive, possessive feelings were not the only change in my brain. Before being pregnant, I was by no means an anxious person. My main hobbies included rock climbing and free skiing and, the winter before getting pregnant, I had both skied the mythical Bec des Rosses – the notoriously steep and rocky mountain hosting Verbier Xtreme – and been caught in an avalanche. Less than a year later, I was standing in front of our house with our new-born in a pram in an advanced state of panic; I was scared out of my wits that I would lose grip of the pram, and it would start rolling down the hill. I broke down in tears and refused to take another step. We all had to go back into the house, and I refused to leave it until my husband had organized a rope for the pram that I could tie around my wrist.

This change of behaviour is quite common among women and can easily be explained from an evolutionary perspective. We have already learnt how the foetus controls the mother's body to make sure it receives all the nutrition and oxygen it needs. But the foetus also needs the mother to keep protecting it once outside of the body. And it actually has a mean to do that: it can modify her brain while still in her womb. Brain plasticity is the brain's capacity of making new neuronal connections. This means that the amount of grey matter can change in different regions, and connections between regions can strengthen or weaken depending on how they are used. Hormones influence how this remodelling is performed. During pregnancy, the hormonal levels of both oestrogen and progesterone are very high, which alters the anatomy of the brain. These modifications can be detected up to two years after giving birth.[546] The regions that are involved in perceiving and understanding the emotions of others are especially affected; it helps the mother connect to the baby.[547, 548] The increase in oestrogen and progesterone also activates the regions that evaluate risks[549] so no wonder many women, like myself, changes fundamentally in how they evaluate risk immediately after giving birth.

In her book *The Philosophical Baby*, the developmental psychologist Alison Gopnik says, "We don't care for our babies because we love them, we come to love them because we care for them." This is more than just philosophy and has a physiological explanation. Once the baby is out, the mother's brain keeps changing, but now mainly under the influence of oxytocin and dopamine. These two hormones belong to the body's reward system: hence, the more you care for your baby, the more you feel rewarded. The oxytocin is produced in the hypothalamus as a result of a nerve stimuli. It not only creates the contractions of the uterus and stimulates the breastfeeding, but it also acts as a chemical messenger and has been shown to have a pivotal role in human behaviours including sexual arousal, recognition, trust, anxiety and mother-infant bonding. As a result, oxytocin has been nicknamed the 'love hormone' or the 'cuddle chemical'.[550] Since many of the regions of the maternal brain are rich in receptors for oxytocin, they are more strongly activated by this hormone.[551]

These structural changes induced by oxytocin and dopamine can be seen in Magnetic Resonance Images (MRIs) of women's brains during the first four months after birth. The regions of the brain that are involved in maternal motivation and emotional reward become larger during this time period and so do the regions that process sensory input, such as the touch of your baby, and empathy.[552] But all those postpartum alterations are not something that happens only to biological mothers. Indeed, scientists have found that fathers and foster mothers by spending time with children may get the same amount of oxytocin and dopamine. So, going through a pregnancy is not necessary, neither is it a guarantee, for those changes to occur.[553]

Given the profound upheavals the brain goes through during pregnancy and postpartum, it is not surprising that the transition to motherhood is a delicate period and can induce a higher risk of mental suffering. At least 1 in 10 new mothers has difficulty in taking care of and enjoying her baby.

Bad-box

Have you ever had one of those evil fantasies where you imagined what would happen if you were 'breaking bad'? I sometimes do, as when someone gets a beautiful but fragile gift, and I wonder: "What would happen if I just smashed all these porcelain plates that she just got for her birthday?" Well, the same kind of thoughts about hurting your baby is also very common. As many as 30% of healthy parents have reported having contemplated the possibility of harming their new-borns. In the weeks before delivery, 95% of mothers and 80% of fathers have such disturbing thoughts.[554] But like some sexual fantasies, they belong to your inner world and do not lead to any acting out (except in a few pathological situations). These thoughts create a lot of guilt and anxiety though, and parents tend to overcompensate with behaviour such constantly checking on their baby. If this is the typical conduct of obsessive-compulsive people, bear in mind that an elevated level of anxiety and distress is normal in this situation. In times when looking after a new-born was a treacherous and all-consuming task, a period of high alert may have helped the parents protect their babies from environmental harm. The mothers who were more careful with their baby were more likely to keep her or him alive, which would account for this residual obsessive-compulsive behaviour which is only natural.

Despite all the changes happening to the brain, there are no signs of less good cognitive behaviour or memory loss, so nothing to support the theory of the 'mummy brain'.

Y-box

The mother's brain is not the only one to change when becoming a parent: the father's brain does too, as well as his levels of hormones.

On the hormonal side, testosterone levels in men decrease when they become fathers.[555] The lower testosterone level helps them release more oxytocin and – by association – dopamine, which facilitates the bonding with the baby.[556] Researchers have also shown the correlation between the father's testosterone level during the mother's pregnancy and his subsequent involvement with the baby and household chores.[557] On the cerebral side, men experience the same increase in regions involved in empathy and understanding the children's needs.[558] But whereas the mothers' regions for nurturing and detecting risk were further evolved, the regions for goal setting, planning and problem solving became more active in men.[559,560] Other skills that we normally associate with the role of the mother, such as the capacity to recognize their baby's cry is also an acquired skill that improves the more time is spent with the baby. This means that it is something fathers and other caregivers can develop.[561]

Baby blues – close to unavoidable

Feeling temporarily down and very emotional just after giving birth is completely normal. Approximately 80% of all women suffer from the 'baby blues' during the first couple of weeks.[562] The brain is not the culprit here, but rather the hormones, and more precisely their sudden and vertiginous drop. In the chapter about premenstrual syndrome, we tackled the problem of the effect of a hormone withdrawal. The same happens here. But worse. Oestrogen acts as your feel-good hormone by stimulating the production of serotonin in your brain. During the whole pregnancy, the oestrogen levels are very high but, just after giving birth, when the placenta is lost (which was the producer of oestrogen), the oestrogen quickly plummets from about a hundred times the normal concentration to very low within a couple of days.[563, 564] This sudden drop in oestrogen leads not only to a smaller production of serotonin, but also to the increase of an enzyme (enzymes are molecules that speed up a reaction) that metabolizes serotonin. It means that the mother almost entirely loses all her happy hormones overnight so it is no wonder that the mood becomes a severe rollercoaster just after birth.[565] For most women this is a temporary change, but for women with postpartum depression, the ordeal can go on for weeks, months or even longer with enzyme concentration staying high[566] and the serotonin one staying low.

Postpartum depression

When a friend of mine gave birth to her twins at the age of 40, she was thrown into a deep depression. Soon after her return from hospital – and with two newborns to look after – she quickly lost control. By the time the twins were 2 weeks old, she was so sleep-deprived that she could not function anymore. She had no energy left to feel anything for them and was just crying all the time. Despite hiring a nurse that looked after the twins every second night, there was no way she could get back on her feet.

She said that she did not have any happy moment with her children during the first year and was unable to develop feelings of love towards them. She felt guilty, she thought she was a horrible mother, and she often regretted having children altogether. This turned into anger against the children.

So, what happened? She fell in a vicious circle and was brought down by a mixture of hormonal imbalance, complicated context and sleep deprivation. We have just seen about the 'baby blues' – a misnomer really. Add short and disrupted sleep: being sleep deprived may interrupt the production of serotonin and disturb your feeling of wellbeing. Add stress, which may lead to high cortisol levels, which in turn will stimulate the very enzyme that breaks down serotonin thereby reducing the serotonin levels even further.[567] My friends were up against a very powerful and grim adversary: postpartum depression.

The loss of progesterone can also play a role in the onset of the depression. As the body has become used to very high levels of progesterone during the pregnancy, the receptors to which it binds have been downregulated – meaning that they need more progesterone to trigger the same reaction – which can lead to a backlash. In the beginning of the pregnancy, the receptors are not yet downregulated which is why during the first months the increasing progesterone can cause a lot of symptoms such as sleepiness. After that, they get desensitized, which may lead to depression when the retraction of hormones happens.

It took my friend four years to get out of her postpartum depression. While the healthcare system was of no help, she eventually turned her life around by herself. Sport came to the rescue. She had heard that running could help against depression, so she decided to give it a go. Since she was overweight and out of shape, it was not an easy challenge. She began with a 1-minute run. Slowly but steadily, she improved, on all levels: she was able to run for longer times, she was sleeping again and was beginning to feel better. The more she ran, the better she felt and eventually she managed to put the depression behind her and enjoy her life and her family.

It can be very heavy to get out of a depression. Generally, cases are solved within the first 3–6 months, but many women still experience symptoms after three years.[568] There is no magic formula but there are options and with the right treatment you can find relief.[569] On the Mayo Clinic homepage they recommend you contact to your doctor if your symptoms do not go away within two weeks, if your depression makes it hard for you to care for your baby, and if you have thought of hurting yourself or the baby.[570]

Men-box

Postpartum depression is not unique for mothers; new fathers are also at risk. In 2016, a meta-analysis of research found that 8% of men experience postnatal depression.[571] However, other academics suspect that the screening tools for detecting depression in women are less reliable when applied to men, suggesting that the real figure is much higher.[572] Elia Psouni from Lund University in Sweden, says that 22% of male respondents in the study had experienced postnatal depression.[573]

Postpartum anxiety

As previously mentioned, some anxiety is normal with a new-born baby. I can't tell you how many times I pulled my babies out of their cribs in complete panic – when they were quietly sleeping – because I thought they had stopped breathing. Actually, I can tell it exactly, because it was such a trauma that it is imprinted in my memory: three times with my firstborn and once with my second.

However, there is a threshold where this anxiety is no longer considered normal. This is when you are constantly, or near constantly, experiencing worry that cannot be eased and strong feelings of dread of things you fear will happen, racing thoughts and sleep deprivation.[574] The anxiety can also give you physical symptoms such as fatigue, heart palpitations and hyperventilation. When it goes so far, doctors talk about postpartum anxiety. This condition is much less known than postpartum depression but similarly quite common. A large study from 2014 showed that 18% of new mothers reported anxiety symptoms.[575]

'Failing' birth – post-traumatic stress

I have already mentioned how my first delivery had been such an unpleasant – and understatement really – event. Since it felt wrong to keep moaning about this horrible moment – because I loved my daughter so much and because 'both mother and baby were doing very well' – I pushed these memories away. When I got pregnant with my second child, those feelings resurfaced: I was absolutely terrified and distraught at the mere thought of going through that immense pain again. I heard about a midwife who had specialized in this kind of situations, and I went to see her.

She sat me in an armchair, gave me a packet of paper tissues and said: "Now tell me, how was it really?" I just cried and cried for an hour and used up my supply of tissues. When the hour was over, I was a new person. One week later I came back, and my fear was entirely gone. I was all right until the contractions for my second birth started. But that time, it all went so fast that I hardly had time to worry at all and giving birth so quickly – yet as painful – healed me from the first trauma.

I think that a lot of my trauma was linked to the feeling that I had 'failed'. I had this silly notion that I would be able to give birth 'naturally' without the help of an epidural, and since I could not handle the pain and needed the help, it felt like a failure. Having a safe space to share my story and having someone acknowledge its brutality and banality was what I needed to embark the next adventure with more serenity.

Many women experience a distressing delivery and feel traumatised by childbirth. Considering how I felt after what was considered to be an 'easy' birth, I am not surprised that up to 4% of all women experience posttraumatic stress disorder (PTSD) as a consequence of pregnancy and birth.[576] One study from 2003 found that around a third of mothers who experienced a 'traumatic delivery' – that is a delivery that involved complications, the use of instruments to assist delivery, or near death – developed PTSD.[577]

That being said, women are not the only ones suffering the aftermath of a traumatizing birth. Partners are also profoundly affected of seeing their loved ones go through such an experience without being able to help and for some it can have a long-lasting impact on their mental health.[578]

Breastfeeding

Breastfeeding is a burning issue with the potential to polarize and antago-nise a room at the speed of light. In reality, it is an irreconcilable experience as well, being wonderful when it works, and nightmarish when it does not. I have had the opportunity of experiencing both scenarios. With my firstborn, it was a fantastic experience. Yes, it did hurt a bit in the beginning, and it took us a while before my daughter and I got the hang of it. But from then on, I enjoyed long, marvellous and relaxing moments of close connection with my daughter. Things went on quite differently with my second child, my son. He would voraciously empty both breasts within 5 minutes while constantly hitting me out of frustra-tion for not getting food fast enough. This crazy feeding-frenzy resulted in me having one breast infection after the other, and every second week I would be in bed, shivering with fever as a result.

All research agrees that breastfeeding – when it works – is the best for your child. I would never argue against that, but the alternative – bottle feeding – will not harm your baby. Every mother wants what is best for her baby, but sometimes you must just accept that the second best is good enough, and that your own wellbeing is also important for your baby. With the quality of today's baby formulas and if you are living in a country with access to clean water, you have nothing to worry about, should you be unable to, or decide not to, breastfeed your child.[579] No matter what you decide, in the beginning it is a tricky business to make it work. In interviews with almost 3000 women, 92% of participants reported one or more concerns at the third day after giving birth, 52% of the mothers had 'technical' problems when breastfeeding 44% reported pain and 40% had worries concerning the milk quantity.[580] But if you manage to make it past the initial hurdles, it can be amazing.

Now, how do we turn into industrial dairy cows? The process already starts during the pregnancy, when the amounts of oestrogen increase and prompt the growth and branching of the breasts' ductal system. The breasts increase in size by adding more connective tissue and fat. The progesterone stimulates the milk producing glands (lobuli) and the development of small storage sacs (alveoli) for the milk in the breast. All these changes happen during pregnancy, which means that the size of breast that you have before pregnancy does not influence your capacity of breastfeeding. The initial size of breasts is only determined by the amount of fat storage. During pregnancy, another hormone also increases, the prolactin. As its name implies, the prolactin stimulates the production of milk in the breasts, a stimulation inhibited by the high levels of oestrogen and proges-terone. But as soon as their levels drop, with the delivery of the placenta, the prolactin has free reign to stimulate the milk production. It will take a couple of days, and up to a week, to start producing the real milk.[581] Indeed, while the baby is sucking to stimulate breastfeeding, the breasts only produce colostrum at first, a low-fat, high-protein, yellowish 'milk'. The colostrum is a precious substance, full of antibodies, leukocytes, stem cells and a myriad of other things that pro-tect the baby's digestive tract. Since the new-born's stomach is tiny with little room for food, the colostrum is produced in very small, concentrated quantities. Besides providing nutrition and protecting the baby from infections, the colostrum is a laxative, which helps the new-born to pass his or her first stools.[582] The production of colostrum will continue for 5 days to 2 weeks and thereafter gradually converse to mature milk after 4–6 weeks.[583]

Every time the baby is breastfed, nerve signals are sent from the mother's nipple to the hypothalamus, which stimulate the production of prolactin and hence of milk. As long as you keep breastfeeding – and up to an hour after that – this stimulus will continue, and the breasts will continue to produce milk.

If prolactin is the hormone producing the milk, it is not the one propelling the milk out of the breasts into the baby's mouth. That is the work of another hormone, one we know very well by now, the oxytocin. When the baby feeds, the nerve stimuli from the nipples does not only stimulate prolactin from the hypo-thalamus but also oxytocin. The oxytocin passes through the blood down to both breasts, where it squeezes the milk out of the storage sacs into the ducts. The process is quite quick (or maybe slow from a famished perspective?): from the moment the baby starts to suckle, it takes up to a minute until the oxytocin acts and starts ejecting the milk. However, for some women, the mere sound of the baby crying can create enough oxytocin to start ejecting milk.[584]

It was previously believed that the baby got the milk out by squeezing the nipple. Yes, quite like when you milk a cow. This is, however, not how it happens. Instead, the baby creates a vacuum with its mouth and tongue, which brings the milk from the ducts and out of the breast. Nipple pain is caused by wounds but also by an excess of vacuum which makes the nipples very soar.

Squirt-box

Breastfeeding can be wonderful, painful, fascinating, boring, strenuous or magical. And also, quite funny. Indeed, when the baby loses its grip, the milk can keep being ejected like a jet. The expression of my son – the couple of times it happened to him – with his face sprayed with milk was quite price-less. The milk can squirt far, even without a baby on the receiving end. In a very unscientific survey on a motherhood forum some claimed to be able to squirt milk up to 20 feet, which is about 6 metres. My personal record is a more modest 2 metres.

The baby will also produce oxytocin while breastfeeding, mainly due to the skin-to-skin contact but possibly also due to the act of sucking. The released oxytocin increases the bonding between the mother and the baby and has a calming effect on both parts. Even without the breastfeeding itself, the oxytocin will be produced simply by caring for the baby and especially with skin-to-skin contact.[585] This means that it is open for all caretakers to create strong and long-lasting bonds with the baby, it is not a unique privilege of a mother. Researchers from Emory University have also shown that giving fathers extra boosts of oxytocin can help them bond stronger with their babies.[586]

My first days of breastfeeding were extremely painful, and not only in the breasts. Something I did not expect was the pain of my uterus contracting. Indeed, the same oxytocin produced during breastfeeding also stimulates the contraction of the uterus. Although terribly painful, the contractions are bene-ficial for the body for they help the uterus regain its initial form and the body recover quicker.[587]

The breast milk contains everything the baby needs to grow, with plenty of proteins, different sugars, fats and vitamins. Its content evolves as the baby grows and it always has the perfect composition for the current needs. It also contains a lot of antibodies that protects the baby against diarrhoea, pneumonia and ear infections.[588] The mother can produce up to 1.5 litres of milk every day, which requires a lot of energy and nutrients. It saps heavily the calcium and phosphates stocks of the mother, which is often taken from her bones unless she drinks large quantities of milk herself.[589] Breastfeeding uses 25% of the body's energy, whereas the brain uses 20%.[590]

Back to fertility

Do you remember the hypothalamus, the gland in charge of the hormonal balance and the one that coordinates the menstrual cycle? When you breastfeed regularly, the hypothalamus halts its secretion of gonadotropin-releasing hormone, either directly by the nerve signals or by the prolactin. Without this hormone the menstrual cycle stops so while a woman is breastfeeding fulltime, it is therefore common that she has no menstrual bleeding. However, as soon as she starts having longer breaks between the feeding sessions, and the gap becomes long enough (which length is very individual) for gonadotropin releasing hormone to be produced. When the amounts are sufficient enough, the menstrual cycle will start again. But remember that the ovulation and fertile phase occur before the first menstrual bleeding, making it difficult to know when the fertility has been restored.[591] To rely on breastfeeding as contraception can therefore be dangerous business.

How it all starts – The first transformation

In some cultures, the first menstrual period is celebrated as an important milestone. In Japan, the whole family gathers to eat a traditional meal called sekihan. Made of sticky rice and a type of red beans, its colour symbolizes happiness and celebration. Girls from some Native American tribes receive gifts while specific rites are performed in their honour to celebrate their 'coming of age'. So is the case in Fiji, where the entry into womanhood is celebrated with a feast within the family.[592] My own celebration was relatively humble. I remember sneaking to my mum, whispering: "I think it happened." In return I got a pat on the back and a pad. That was it.

Still, it was a much better experience than most girls around the globe, who instead of being celebrated for having their periods need to deal with heavy social stigma. The most ludicrous stories circulate. They are told that if they look at someone while having their period, they will make them sick. If they cook or touch food, it will rot. If they go swimming, they will be eaten by sharks.[593] I am not sure whether my daughter will allow me to celebrate her the day she will get her first period. We could have a family night out at her favourite restaurant, or a secret ritual just the two of us. If she chooses not to make a big deal out of it, I have already shown her, where the pads are and told her that if it happens in school, she is allowed to say she has a headache and come home. After all, it is not every day you will get your first period. What is important for me is that she realises that this is something to be proud of and make sure she is aware of all the positive things it symbolizes: becoming a woman, being healthy and being able to create life if you chose to.

The first period is referred to as menarche, which comes from Greek: *menos* meaning month and *arkhe* which means beginning. It is not the first sign that our daughters are growing up, since menarche happens years after the first signs of puberty appear. A friend of my daughter was only ten years old when her mother took her to the doctor in total panic. She had felt a strange lump under one of her daughter's nipples that had not been there before. Convinced it was something potentially lethal, she came back from the doctor embarrassed but relieved. No, it was not a mortal tumour but the first signs of her daughter entering puberty and her breasts starting to develop. It had begun much earlier than she'd expected and, since it was just one of the nipples, she did not make the link to puberty.

Another friend of mine was quite shocked when her daughter got pubic hair already at the age of eight. The daughter was still such a young child. Another typical sign of growing up is that suddenly your daughter goes from being a kid who only needs to bathe once a week, to someone you need to force into the shower at a much higher frequency since the hair getting oily, and the body odour becomes much stronger. These changes are the precursors of puberty, the adrenarche.

Adrenarche is the activation of certain hormones built by the adrenal glands. Androgens are synthesized in the testes, the ovaries, and the adrenal glands and the best known one is testosterone. Besides androgens the adrenal glands produce hormones like adrenaline, the steroids aldosterone and cortisol.[594, 595] Androgens are the precursors to oestrogens in both men and women. In women the androgens will be turned into oestrogen inside the ovarian follicles. The adrenal glands are situated just above the kidneys and their job is, among others, to send out the stress signals cortisol. When you are about six to eight years old, these glands start producing androgens which will push the development of pubic hair, body odour, skin oiliness, and acne. These symptoms are often the first signs of the approaching puberty and they advertise the beginning of a hormonal storm.

Puberty

The hormonal awakening ("She's alive!")

The years preceding puberty are quite a remarkable period. Previously, the girls in our neighbourhood grew up, and I saw them become slightly bigger versions of their five-year-old selves. But suddenly and with a bang – as if someone had waved a wand and casted a transforming spell – you cannot recognize them anymore. They have grown, seemingly, a decimetre or more, changed shape and their facial features are completely different. They have undergone a complete transformation almost overnight. The whole process starts with the adrenal glands activating certain hormones and opening the door for puberty.

As we know from the menstrual cycle, the hypothalamus – a part of the brain – acts as the main orchestrator of the hormonal coordination in the body.

So, it comes as no surprise that the hypothalamus is the headquarter for this revolution and where puberty is initiated. During childhood, the hypothalamus is partly inhibited but in the beginning of puberty it starts reaching its full function. Researchers haven't understood precisely why the inhibition of the hypothalamus suddenly stops, but they know of one of the factors involved in the process, is the hormone leptin.[596] Leptin is produced in fat cells. To produce enough leptin for the hypothalamus to become fully activated, you must have enough body fat. Hence, in order for menstruation to start for the first time, it is necessary to have at least 17% body fat.[597] It explains why puberty happens earlier in girls with a higher BMI.

Orchestration-box

All the hormones needed to coordinate the menstrual cycle are produced in the hypothalamus and its close companion, the anterior pituitary gland. This is a small, pea-sized gland situated just below the hypothalamus. The hypothalamus starts producing GnHR which triggers the anterior pituitary gland to secrete FSH and LH. This happens for both boys and girls, even though girls are normally a few years earlier.

In girls, the hormones released by the anterior pituitary gland will stimulate the follicles waiting in the ovaries to mature. In the beginning, it is the amount of LH that increases most, which stimulates the ovaries to produce testosterone. Thanks to the FSH, which also increases but at a lower level, the testosterone is turned into oestrogen inside the growing follicles. The remaining testosterone drives changes that are also linked to puberty, such as pubic hair and acne.

Each girl will develop in her own unique way. Among factors that influence in which unique way a girl will develop is the speed at which the testosterone is converted to oestrogen. This will affect the amount of typical puberty traits, such as acne and pubic hair she will get.

During puberty, the oestrogen level continuously rises, which triggers all the physical changes happening both within and outside of the body. Since testosterone is turned into oestrogen, more oestrogen means less testosterone, and this effect gets even stronger as oestrogen produces sex hormone binding globulin. This is a molecule which makes sure that the testosterone cannot circulate around freely. With less free testosterone, some of the symptoms, such as acne, disappear quicker for girls than for boys.

Loop-box

The overall surge of hormones will continue until the feedback loop with the hypothalamus is established, and the hypothalamus takes over the full control. The hormones then stabilize at the levels typical for the menstrual cycle. Whereas low levels of progesterone were produced in the adrenal glands beforehand, it will now only be produced when the menstrual cycle is fully established and as a consequence of ovulation.

Bodily changes – the metamorphosis

With a daughter approaching her teens, I should have had the perfect opportunity to get first-hand information about a girl's life at the early stages of puberty. But every time I tried to have a conversation with her about this topic, she told me that I was disgusting, that I should get out of her room and that I have just ruined her day. So, I had to dig into my own memories instead.

I must admit that these memories were hazy and felt very distant, and that I had to work very hard to dig them up. I vaguely remember of being one of the early girls to grow pubic hair. I found that very embarrassing in the shower after gym class and I did my best to hide it. Breasts are more visible to boys and I was very grateful not to be a precursor in that aspect. A classmate of mine, Linda, was the first to develop breasts, with all the boys making fun of her. They would use both hands to pull out their T-shirts mimicking breasts and chant "Here comes Linda!"[598] It must have been a terrible way to enter puberty and looking back I feel deeply sorry for her.

Although I mostly tried to ignore and hide the changes that were happening to me, my best friend at the time reacted very differently. She was super curious about the changes she was going through and always wanted to compare the little lumps that started to develop behind our nipples. I felt very uncomfortable with that. I was a bit of a tomboy at that age and turning into a woman was the last thing I was interested in. I guess a little bit like my daughter now.

The Tanner stages

Even though each individual has a different puberty timetable, there are some general phases most girls will go through in their physical development. These are the Tanner stages or the sexual maturity rating.[599] There are five diffe-rent Tanner stages, describing the characteristic features of each phase from around 8–15 years old, when puberty ends.

Tanner stage 1: The hormonal activation of the hypothalamus and the hormonal interplay described above is referred to as Tanner stage 1. It normally happens sometime after the eighth birthday. At this stage, no external signs are perceptible yet.

Tanner stage 2: From childhood to puberty, oestrogen increases almost 20-fold. The bodily changes that we associate with the female shape are all triggered by the oestrogen and the impact of these changes starts to be visible in Tanner stage 2. It normally happens between ages 9 and 11. As previously mentioned, girls with higher BMI have an earlier onset of puberty.

It is in this stage that the first signs of breasts – botanically dubbed 'buds' – start to form beneath the nipple. The buds may be itchy or tender, and the dark area around the nipple will expand. A light tuft of hair also become visible. As seen in the examples above, pubic hair sometimes starts growing before the breasts do, and it is also very common that one breast develops much faster than the other. It can take 6 months for the other breast bud to become noticeable,[600] a period that can feel interminable and be anxiety-inducing, as made apparent by the many websites answering the question "Why are my breasts different sizes?" or "Is one breast being bigger than the other a sign of cancer?"

Just after the breast buds appear – but before the first menses –a growth spurt starts. This sudden increase in size typically takes place when girls are 10–15 years old, which is about 2.5 years earlier than for boys. The growth spurt will last for about two to three years. After that, oestrogen has reached such high levels that it slows down the growth. Another effect of the oestrogen is that it intensifies the sensitivity to insulin. Insulin impacts the amount of body fat and lean muscle a person can develop, explaining the differences in muscle mass between men and women.[601]

On the inside, oestrogen stimulates all the transformations necessary to turn us into baby-making machines. Under the influence of oestrogen, the ovaries, fallopian tubes, uterus and vagina increase several times in size. The vulva grows and modifies its shape, and the inner labia gets more wrinkled. The hymen – which is not a membrane but rather a wrinkled mucous membranous ring at the entrance of the vagina – becomes more elastic and thickens.

Sexual maturation means that the body gets ready to have intercourse. The oestrogen changes the inner wall of the vagina, making it considerably more resistant to trauma and infection than before puberty.[602] The bacterial environment of the vagina is altered and, approximately one year before the first period, the vaginal pH becomes more acidic. The vaginal pH matters as it is a sign of health and an acidic vaginal environment is protective. It creates a barrier that prevents unhealthy bacteria and yeast from multiplying too quickly and causing infection.[603] Vaginal discharge is often a sign that the first period will soon come.[604] Vaginal discharge is a way of the body to discharge cells and liquids. It is made of fluids from the uterus, cervix and vagina. It is normally clear and its colour and if healthy, it should not smell bad. Once the menstrual cycle is established it will change in consistency dependent on the cycle-phase.

Since the capacity to have sex needs to be paired with the capacity of enjoying it, the clitoris also starts growing in this phase. In this case, it is a consequence of the increased amounts of testosterone and not oestrogen. This may not be surprising since the penis is the male counterpart of the clitoris. Both have originally evolved from the same structures, and both develop at puberty thanks to testosterone.

Tanner stage 3: In Tanner stage 3, the physical changes become more visible on the outside. The breast buds continue to grow and expand, the pubic hair gets thicker and curlier, hair is forming under the armpits.

Bush-box

The evolutionary purpose of pubic hair – and the reason why it does not have the same texture as the hair on the head – is most likely a way to signal sexual maturation and to protect our nether regions.

The curly, denser type of hair helps reducing friction on our genitals, both during sex and other activities. Hair can also keep the genitals warm which make us hornier. Pubic hair is beneficial because the hair follicles produce an oily substance that can prevent production of bacteria and protect against certain infections.[605] It is interesting that our hairy cousins, the gorillas and chimpanzees, are actually less hairy in the genital regions, as a curious blogger for Scientific American pointed out.[606]

Is it then a good idea to shave the pubic hair? There are no hygienic reasons for shaving and its sole purpose is aesthetics. It is not a new trend, but something we have been doing for centuries. There are benefits of not removing all of it, but as long as you do not experience any negative effects of the removal, even physicians will leave the choice to you.

In this Tanner stage, fat is deposited on hips and thighs, and this is the period where the growth really takes off and girls suddenly shoot up.

In this stage acne may appear. I remember suffering from acne when I was in puberty, but we talk about a few pimples here and there. Even though I felt that they were ruining my life, it was nothing compared to the acne invasion sustained by my younger brothers. As a rule of thumb, girls get less acne in puberty than boys. The acne is triggered by testosterone and once enough oestrogen is produced from testosterone, the oestrogen will suppress the activity of the glands that produce oily substances.[607]

In puberty, the voice changes and develops differently for boys and girls. The oestrogen makes the voice box smaller and the vocal cords shorter, which gives the girls a higher-pitched voice than boys.

Tanner stage 4: In this stage, the breasts get a fuller shape, the pubic hair gets even thicker, while height growth slows down a little. This is when the first period might come, the menarche, but it will take approximately 18 months from the first period until the cycle becomes regular.

For about six years, 90% of 'normal' 12–14-year old girls don't have a regular ovulation during their menstrual cycles.[608] The typical age of the first menstruation is between 11 and 16 years old, with an average of 13 years.

Tanner stage 5: This is the last development stage and the moment you can say that you are fully grown. In girls, this stage usually happens around the age of 15. Now, the breasts reach approximately adult size and shape, the periods become regular, the reproductive organs and genitals are fully developed, the hips, thighs, and buttocks get fuller. The girls reach adult height one to two years after their first period. Other effects of the oestrogen are that the pelvis and hips widen. By Tanner stage 5, the amount of body fat has amplified to the amount of a grown woman. Oestrogen also causes the lips to grow and get a redder colour.

All set (or not)-box

When are you ready to have children: Having got your period is in no way an indication that the body is ready to have children. The body still needs to mature properly. Early childbearing can increase risks for new-borns as well as young mothers. Among girls aged 15–19, pregnancy and childbirth complications are the leading cause of death globally.[609]

Anatomically, there is a higher risk of preterm delivery as the cervix is still not fully developed. Immaturity of the pelvic bone and the birth canal is also an additional risk, both for the mother and the baby, during delivery.[610] When the body is not fully developed, the risk of anaemia (lack of iron) is three times higher, which can lead to low birth weight and preterm labour. Babies born to mothers under 20 years old face higher risks of low birth weight, preterm delivery and severe neonatal conditions.[611] Young mothers aged 10–19 years also face a much higher risk of eclampsia and various infections compared to mothers aged 20–24 years.[612]

On top of that, there is also a strong social risk with teenage births: risk of poverty, dropping out of school and living in an abusive relationship.[613]

So even though it might be technically possible to get children earlier, it would be wise to wait until you are at least 20. Your best chances of having a successful pregnancy with a healthy outcome for you and your baby is 20–35. The upper limit is not a 'hard limit', but you are statistically more likely to have a successful pregnancy during this age-span. You can read more about this in the fertility chapter.

Mood and brain – teenager on the verge of blowing a fuse

A friend of mine told me that she could completely freak out as a teenager and was not able control her outbursts of anger. Once, she broke a window of the family's house, because she was so upset that her key did not fit. The story does not tell if she used the right key or if she was entirely sober, but the anger was totally disproportionate.

On a personal level, puberty was a much calmer period – all windows remained intact – and the only epic conflict I can remember occurred when I was around 12 years old. I guess it was a very Swedish type of conflict. During a long tirade of Swedish profanities, I informed my father that I would never again go cross-country skiing with him. I might also have thrown a ski at him.

The emotional storms teenagers experience have many reasons and can be linked to the changing hormones but also to a complete remodelling of the brain and neurotransmitters. These emotional rollercoasters are at the strongest in the early teenage years and calm down over the years. Girls tend to have stronger mood swings than boys, which might not be surprising considering all the different hormones that they need to deal with on top of the testosterone.[614]

In the chapter about the brain and the hormones, we discussed the influence of sex hormones on mood and wellbeing. When a girl reaches puberty, all this hormonal impact is felt for the very first time. It is no wonder that this may be overwhelming and difficult to handle for a young teen, especially since hormones are not the only ones to affect you. Neurotransmitters also play an important role in mood and behaviour in this phase. Dopamine and serotonin influence brain areas that control how we react to emotions, and the ability to experience pleasure and pain. During puberty, their levels decrease, which results in mood swings and makes it more difficult to control impulses.[615]

There is a massive remodelling of the brain during puberty. The brain of teenage girls reaches its final size already around the age of 11, but this does in no way mean that it has finished developing. It will continue to change until their mid-twenties.[616] The puberty remodelling consists of the development of faster pathways in the brain and the multiplication of neurons. The improvement of these pathways is influenced by the sex hormones, which explains some of the differences that can be observed between the average man and the average woman later on. For full disclosure of those differences, please go to the chapter on how hormones impact the brain (page 82).

The special ability many teenagers have to start conflicts is most probably linked to the late development of the part of the brain that directs the capacity to exercise good judgment. This is the prefrontal cortex, the part of the brain lying just behind the forehead. The prefrontal cortex is responsible for cognitive analysis, abstract thinking, and the moderation of behaviour in social situations. This part of the brain is one of the last parts to mature and, in some people, it happens as late as the age of 25 (and in some others, it can feel as if it never occurred at all, but that is another problem entirely). The area is completely re-modelled during puberty which impacts both motivation and emotions. An immature prefrontal cortex can manifest itself in many ways, including mood swings, conflict with authority and risk taking.[617] Adolescents trying to interpret the world react with the emotional region of their brain instead of relying on their prefrontal cortex. The result is a much more impulsive reaction than most adults would consider logical and appropriate. [618]

But beyond the outbursts and the conflicts that can characterize the every-day life with a teenager, it is also a remarkable period to connect and have amazing conversations. The major brain remodelling creates new improved con-nectivity which allows for multitasking, better ability to solve problems, and the capacity to process complex information. This makes puberty a particularly good period to develop talents and lifelong interests. It is, however, also a vul-nerable phase, where trauma, chronic stress, drug abuse and sedentary lifestyles may have a particularly negative and long-lasting impact for the future.[619]

The issue of sleep – a grudge against the sandman?

The sun rises, everybody wakes up and gets out of bed. Showers are taken, coffee drunk. The household gets to their activities, studying, tiding, working. The whole household? Well no. The teenager is still fast asleep. But the parents might want to stop the grumbles and moans, for recent research have shown that it is not a matter of laziness or bad habit. Indeed, it is attributable to a hormonal change that is beyond their control. It is due to another hormone that changes during puberty, namely melatonin. Melatonin is the hormone that regulates the circadian rhythm, the rhythm of day and night, and makes us sleepy in the evening. It is to a large extent regulated by light, but there is now more and more evidence that this light sensitivity is altered in young people during puberty. Hence, the release of melatonin is delayed in the evening, meaning that they don't switch to 'go night' as quickly as adult.[620, 621] The exact mechanism is unclear but some sympathy from the parents' side might be warranted if they need to spend a bit more time in bed during the weekends. It is not something they can control, and they still need a lot of sleep for their brains to develop; 9–10 hours per night is recommended. More and more voices are now being heard to advocate for schools starting later, since this seems to be a way to provide teenagers with the sleep they desperately need.[622]

An earlier puberty – against the clock

Over the last decades girls are reaching puberty earlier, and the average age is almost a year younger today than it was 40 years ago.[623] This last century, the change is even more striking: in 1820, the average age for puberty was 16.6 years; in 1920, it was 14.6; in 1950, 13.1; 1980, 12.5; and in 2010, it had dropped to 10.5.[624]

What is happening to our girls? Researchers have ventured several explanations. The main one pertains to leptin. As we have seen, this hormone is produced in the fat cells and plays an important role in the start of puberty. So, the fact that we tend to be very well nourished nowadays – and even too well nourished in certain cases – is most likely the main factor.[625] Another explanation could be linked to all the artificial lights we are exposed to. Melatonin, which is the hormone released when we sleep in the dark, is known to suppress other hormones that can trigger puberty. But in the modern environment where it is rarely dark, less melatonin is produced, leaving the field clear to trigger the puberty process.

This theory is supported by the evidence that children living close to the equator, where there is more natural light, have an earlier puberty than children living close to the poles, regardless of their genes.[626]

Interestingly, early puberty has also been linked to not living together with your biological father. Scientists still speculates on the possible mechanisms behind this fact, and the theories go all the way from the threat of being away from your biological father to being triggered by being close to a male that you do not share genes with.[627]

Another culprit of this precocious puberty can be found in the large number of chemicals that we are exposed to through personal care products.[628] Many countries, including the EU, have taken action to ban use of the worst chemicals and they continue to work on the regulations. Consumers can also decrease their exposure by not buying products containing chemicals that might cause harm.[629] The most likely suspects are the endocrine disruptors phthalates, bisphenol-A, and pesticides – but the scientific findings are inconsistent, and more research is needed to be able to draw any conclusions.[630, 631]

We have now learned how the reproductive lives of women kicks off. Once up and running, the average woman will have 400 menstrual bleedings and according to ourworldindata.com, she will have 1.7 pregnancies if she lives in Europe or the United States, and with an average of 2.5 if the whole world is included.[632] After about 30 years of service, the system will then start to slow down to finally shut down about 40 years after it all started. This is the menopause, and we will learn about that in the next chapter.

MENOPAUSE – THE FINAL TRANSFORMATION

Now we have come to the last subject of this book, which also turns out to be the last chapter of a woman's reproductive life. The end of her reproductive life yes, but in no way the end of her life, nor her faculty to enjoy it, seduce partners or have an active sexual life. Menopause is a natural transition, and not a pathology, and although it brings along some discomforts it sometimes tends to be more dreaded than what is called for. We'll come back to that. But let's see first how evolution is assessing menopause.[633]

The grandmother theory

From a raw evolutionary perspective – where the sole purpose of all life is to reproduce – outliving our fertile years doesn't make sense. The conundrum of why humans live beyond their fertility has puzzled researchers for a long time. The American anthropologist Kirsten Hawkes theory from the 80s is still the strongest hypothesis to this day: the grandmother theory.

The grandmother theory suggests that, by helping their daughters raise their grandchildren, the grandmother's genes can survive even after their reproductive age. It would therefore make evolutionary sense for women to live past menopause. Besides humans, few other primates live decades beyond menopause. Most animals reproduce until they die. The only species known to have a substantial number of females living past their reproductive age are four types of whales (beluga, orca, narwhal and short-finned pilot) and one insect, the aphid.

Grandmothers are not only evolutionarily helpful in looking after the offspring. By analysing data from different cultures, it has been found that grandmothers are especially productive in acquiring food. Data from earlier times are scarce but exist from a population of French settlers in Canada between 1608 and 1799, where women with a living mother had twice as many babies as women whose mother had died.[634] The further away the grandmother lived, the less grandchildren would be born or survived. A similar finding has been seen in a pre-industrial Finnish population, where a present maternal grandmother increased survival of the grandchildren after weaning. I will leave it up to you to interpret the following finding but living with the paternal grandmother had the opposite effect on the child's survival.[635]

The inevitable conclusion is that grandmothers are a key factor for success for our entire species. A good reminder to us all to treat them better in society.

Growing older

I am in my early forties and recently, a couple of months after I attended a workshop on perimenopause, I got a shorter menstrual cycle than usual. I had just learned that this is the first sign of my body trying to make the most out of my last reproductive years, so I totally panicked: the countdown to my very last menstruation, the dreaded menopause,[636] had started. Now I would get old and grey. I would never again benefit from the highs that my high oestrogen phases normally give me, no one would ever find me attractive again, and my libido would go down the drain.

To find comfort and information, I did what one does but shouldn't, I searched the internet: there was almost no useful information. Plan B was to call my mum, and her reaction was: "What? What has now gotten into your head, silly?" (I could almost hear her eyes rolling). For her, menopause was not a big deal at all. Instead, her fifties were the time when she finally upgraded her dreams to a new level. Once all of us kids were independent, she went from working as a dentist to being a researcher. She started going to conferences all over the world, meeting interesting people and, first and foremost, started working on a topic that really interested her. For her, this was a positive and inspiring phase of her life.

I then thought about my grandmother. She died at 94 years old and, at her funeral, the priest showed a picture of her when she was 75 and said, "These were her best years." She had become a widow and she could now run her little shop for children' clothes, toys and diapers by herself with the help of her best friend. So, the most interesting part of her life started much later than society imagines. Now life was no longer about being there for everyone else, but actually about her and her own project.

As usual, my first reaction had been far too alarmist. For the vast majority of women, menopause is not a negative experience at all.[637] I managed to get over my fear of getting older, but I still did not get an answer to what will happen to my body in this transition period. Now I know more.

(Un)happy-box

Apparently, we (everyone, not only women) are at our unhappiest at around 47 to 48 years old (47.2 in more economically developed countries and 48.2 in the developing countries).[638] It is difficult to know if this dip in happiness is due to societal and environmental factors or something that is physical. It is interesting that this dip happens just before menopause. Apart from being a challenging time in many people's lives, which is the main hypothesis of the researchers publishing the study, maybe it is the fear of growing old that is depressing for us, and once menopause is over, we realize that it was not so bad after all.

Anyway, if you feel down at the prospect of aging just hang on tight for a short while longer, and you'll soon be 47.3 or 48.3 and things will be better again. (Sorry for the outrageously wrong way of using statistics. If you feel unhappy at 30, don't believe it will be all downhill until 47, this is only an average and as usual it doesn't mean much to you as a person, but it is still interesting.) After 50 many can enjoy a pleasant phase: the children have left home, and their old parents are still well enough to take care of themselves. Now they have time and energy to start new projects and to take an interest in their own life.

The hormonal changes of the late reproductive years

The signs that I experienced – the shortening of my menstrual cycles – were not yet the signs of perimenopause (the period directly preceding menopause) but rather an indication that I have reached my late reproductive years. This means I could still get pregnant, but it would be a lot harder.

As we grow older, the number of follicles in the ovaries gets smaller and smaller. A girl is born with about a million follicles, by puberty this number has dropped to around 250,000 and it will gradually be reduced to around 1000 by menopause. You probably remember that a woman will only ovulate around 400 times so most of the follicles die naturally. During most of the reproductive life, a similar proportion of follicles will be lost every menstrual cycle but after 30, the loss of follicles accelerates. By the time menopause is reached there are only around 1000 follicles left.[639] Oestrogen is produced inside the follicles and as the number of follicles decreases, less and less oestrogen is produced. One of oestrogen's roles is to inhibit FSH. With less oestrogen produced, the amount of FSH therefore increases.[640] As if to make the most out of the few follicles that are left in the ovaries, the increased FSH brings even the less responsive follicles to maturation. The body does everything it can to make sure we keep ovulating. Hence, the menstrual cycle continues but the released eggs are of inferior quality.[641] The increased FSH can also lead to the maturation of more follicles and this is why twin births are more common among older women.

Prediction-box

How can you know if you are approaching the end of egg-stock?

Inside growing follicles, the hormone AMH is produced. As the number of follicles decreases, so does the amount of AMH. As this hormone is easily measurable, this can be a good indicator of the number of follicles that are still left in the ovaries.[642] The amount of AMH starts to decline rapidly around 40.[643] As one of the roles of AMH is to prevent low-quality follicles to mature, a consequence of the declining of AMH-levels is that such follicles can also mature. Low AMH levels are, however, no indicator of your capacity of conceiving naturally. For your current fertility, such tests are of little interest.[644]

The increased FSH also helps the follicles to grow faster, which makes ovulation occur earlier in the cycle and to eggs being slightly smaller. The cycles can become slightly shorter and more irregular, which was what I had observed. For me the difference was so small that I probably would not have noticed it if I had not been tracking my cycle for years. You might not feel any changes in most of your cycles but simply notice some odd behaviour now and then. Up to five years before menopause, more than 90% of the cycles have an ovulation but after that, the number declines quickly and only about 15% of the cycles are ovulatory the year preceding menopause.[645]

Box-on-fire

When my cousin was in her mid-forties, her PMS was going from bad to worse. She got terribly emotional and would suddenly go ballistic and scream at her family. Every month, on the same day of her cycle, she'd consider a divorce. It got so bad that she started organizing her life around her PMS, making sure she would not book any meetings during the period of her PMS. But the period of beserkness began to extend and drag on: from two days, it became a whole week. She went to the doctor and found out that she had very high levels of cortisol without being stressed, which is common during the menopausal transition.[646] This could potentially have been a reason her PMS got so bad, since the cortisol breaks down the serotonin at a faster pace than normal. The high hormonal levels also lead to fatigue and hair loss. Like my cousin, a large number of women get worse PMS and mood-swings in their late thirties and early forties.[647] In my cousins' case, it took several years before she went to the doctor but once she did, she was prescribed oestrogen which helped, and her life got back to normal again.

Perimenopause – a painful transition?

Social representation of aging women is seldom gentle. Often menopausal women are portrayed as some kind of raging Cruella de Vil with her panoply of hot flashes, tantrums and depression. Firstly, this is by no means a fair or accurate representation of menopausal women. Secondly the period those stereotypes allude to is not menopause. Menopause refers in fact to the very last menstruation. The potentially tough period is the few years leading up to menopause, called the perimenopause or climacterium, and with the last cycle, the symptoms disappear.

The average age of entering perimenopause is 47 in Europe, but it can both be much earlier and much later. It starts a few years before menopause and, during this period, the cycles become very irregular. Irregular means that it can differ by more than 7 days between the longest and the shortest cycle. Due to the increased levels of FSH, the follicles are growing faster, and ovulation can happen earlier in this phase.[648] However, as the oestrogen levels are decreasing, it also happens more frequently that ovulation is skipped entirely which means longer periods without bleeding.

Late perimenopause is defined as having cycles lasting longer than 60 days, which means very scarce bleedings. The oestrogen levels are sometimes low and sometimes high and can have a very bumpy pattern. The progesterone is also often low due to the many anovulatory cycles. In this phase, the symptoms we associate with perimenopause can become strong.

Many changes happen to the body in perimenopause. My cousin also got lumps in her breasts which made her – as she and I both belong to the hypochondriac side of the family – jump to the obvious conclusion that she had breast cancer. She went through screening and did biopsies of the tissue, to finally discover that it was simply a consequence of hormonal changes; the lumps soon disappeared again. Such changes are quite common, and so is the breast pain that develops and feels a bit different from the one you can get towards the end of the menstrual cycle. Another change that happens after menopause is that the lower oestrogen levels will make the breast tissue dehydrated and less elastic so they might lose their round shape and begin to sag.[649]

Menopause – the end?

So finally, after many years of good and loyal service, your uterus will shed its lining for the very last time. No more bleeding, no more pads, no more tampons. You will no longer be subject to the ups and downs of the changing hormones. No more premenstrual syndrome. It sounds like liberation and I hope it is the way I will feel about it. This is the definition of menopause, your last menstruation ever.

Menopause occurs when the number of remaining follicles is below 1000, regardless of age. A woman born with a smaller number of ovarian follicles can become menopausal before the age of 45. It happens to about 10% of all women. The average age for menopause varies depending on countries and regions, but is around 51 in Australia and Europe, 49 in the U.S., 48 in Asia and Africa, and 47 in Latin America and the Middle East.[650] It is difficult to know if it is due to ethnicity or other factors as socioeconomic status since there is a study from Nigeria, where the average age of menopause was 52.8.[651]

I had been worried that my sex drive would entirely disappear after menopause since oestrogen will no longer be produced. However, that worry is most likely unfounded. It is not only oestrogen that stops but, as a consequence of no ovulation, the production of progesterone as well. Progesterone is known to make us feel disgusted and often kills sexual desire. Once it is out of the picture, the sex drive can even increase.

Important for lust is also testosterone. Testosterone decreases continuously over a lifetime and is already reduced by almost a half between the age of 20 and 40. However, through menopause, it does not necessarily keep decreasing. In most women, the postmenopausal ovary secretes more testosterone than the pre-menopausal ovary, at least in the first years of the postmenopausal period.[652] With the simultaneous reduction in oestrogen, the amount of sex-hormone binding globulin, which captures the testosterone and makes it inactive, is also reduced. The effect can even lead to freer testosterone through menopause.[653] Even though it is by far not the case in all women, some women actually get an extra push in their sex life when they go through menopause. The majority, however, do seem to get a decrease in sexual desire. To some extent it seems to be linked to decreasing oestrogen levels but also due to symptoms due to vaginal dryness and psychosocial factors.[654] Despite that, receiving both additional oestrogen and testosterone seem to increase libido.

Hot flushes – finally warm?

In her podcast, Michelle Obama talked about going through perimenopause and having hot flushes while on official missions as the first lady. She also revealed that Barack Obama was in no way perturbed by the fact that many women on his staff were going through the same phase and simply suggested to turn on the aircon. But what are those infamous 'hot flushes'? Hot flushes normally occur during the later stage of perimenopause. A hot flush describes symptoms such as the skin suddenly feeling very warm, the face reddening, more or less heavy sweating, or the heart beating faster than usual. It also comes in the same box as night sweats. They normally last 3 or 4 minutes, but they can last as long as an hour. The hot flush itself is due to the dilatation of the blood vessels, which increases the blood flow to the skin. This is what makes the skin so hot during the flush. Normally, this reaction is a strategy to cool the core body temperature when it is too high. It is, however, not the case with hot flushes: here, the core body temperature is normal but decreases afterwards as a consequence of increasing the blood flow to the skin.

Depending on ethnicity, the average number of years that the hot flushes last is quite different. African-American women have the longest median duration with around ten years, followed by Hispanic women with eight years, six years for white and five for Asian women. Since soy is a phytoestrogen, i.e., a plant-based oestrogen, a high soy diet may be the reason why Asian women seem to suffer less from hot flushes.[655] About 85% of all women experience hot flushes, but the severity varies a lot.

Hot-box

Researchers haven't fully understood yet what triggers the hot flush, but they have their eyes on the thermoneutral zone, i.e., the temperature zone where the body does not require any action, it neither needs to be cooled or heated. This zone becomes very narrow due to changes in a chemical called norepinephrine.[656] Norepinephrine acts like a hormone and neurotransmitter in the brain. It is called the 'danger' hormone, and its main role is to mobilize the body and the brain for action. It plays an important role in thermoregulation. Thus, increased levels of norepinephrine probably narrow the thermoneutral zone in postmenopausal women. Serotonin is another likely culprit here. Low oestrogen levels are an additional prerequisite for this to take place. When oestrogen decreases, serotonin also decreases. Decreased serotonin levels lead to increased norepinephrine which disturbs the thermoregulation. All this happens in the hypothalamus.[657] Although the researchers know that the hypothalamus is the main coordinator, they are unsure about the exact chain of events.

It was long believed that hot flushes were linked to lower levels of oestrogen, but it has been proven that there is no relation between oestrogen levels and the severity of symptoms. Oestrogen must play a role somehow because taking oestrogen is efficient against the symptoms. There is a link with the amount of FSH. The higher the FSH, the more symptoms, but it seems to be a complex mix where genetics, cultural factors are involved.[658] Cigarette smoking and high BMI has a negative influence on the flushes, whereas moderate alcohol intake seems to have a positive one.

Mood and brain – is it simply the others?

Menopausal women have the reputation for being cranky, at least according to themselves if one were to listen to my mother and her friends saying that they no more accept stupid decisions from, for instance, employers and colleagues without arguing. So, is it a myth or reality? The internet is full of stories depicting mean menopausal women. Mean would definitely not be a fair judgement and not taking crap from others anymore seem like a rather healthy development to me. That being said, irritability is the most common mood complaint in perimenopause and is reported for 70% of all women.[659]

Here again, oestrogen could be a key factor in mood changes. Indeed, oestrogen directly influences the production of serotonin, the 'feel-good' hormone. With the fluctuation of oestrogen levels, serotonin rockets and plummets, with a potentially dramatic impact on a woman's mood. Without this nice cushioning of happiness chemicals, it is no wonder that the surroundings can become annoying. Mostly due to the lack of serotonin, perimenopause and menopause are periods during which women face a higher risk of depression. Women in perimenopause are up to three times more likely to report depressive symptoms than premenopausal (before any symptoms of perimenopause are shown) women. The strongest predictor is a history of depression, but the changing hormonal levels are also an important factor in the occurrence of depressed mood.[660] Depression in perimenopause can also be linked to dementia and Alzheimer's later on.[661] Neuroscience findings suggest that taking hormonal replacement therapy when you reach menopause could protect you against dementia.[662]

Hormone replacement therapy (HRT) is a treatment prescribed by doctors to women with strong symptoms such as described above. It replaces the 'lost' hormones and can provide oestrogen or testosterone or a mix of both. It has proven beneficial for the overall quality of life in this hectic phase. It can diminish hot flushes, especially during the night, greatly improving sleep. In the same department, it is also helpful since the lower levels of oestrogen and higher levels of FSH can lead to troubles of falling, and staying, asleep.[663]

The brain is influenced through other mechanisms. During perimenopause, the glucose metabolism of the brain decreases by as much as 15% and in menopause it decreases even further due to the disappearing oestrogen that has a large impact on insulin and thereby the metabolism. This can lead to temporary issues with the memory until the brain has adapted to the new state. Despite that, once that short phase is over, studies have shown that women perform just as well in cognitive tests after menopause as they did before.[664]

Other symptoms

It is very common to gain weight during this period: on average, women put on 2 kg in the course of perimenopause. However, this weight gain is not directly related to the changes in hormones but rather to age itself. As we get older our muscle mass decreases and, with less muscles, we burn less calories which can lead to an increase in weight unless we become more active or change our diet. What the hormones do influence is where the new fat deposits appear. As we have learned earlier, the oestrogen promotes the fat to gather around the hips but with less oestrogen, the weight gain tends to settle more around the waist, which is more harmful as it can lead to tissue inflammation and provoke type 2 diabetes.[665]

During the menstrual cycle, the oestrogen played an important role in producing cervical mucus and humidifying the vagina. The low oestrogen production in the menopausal years makes it common with vaginal dryness, irritation and difficult and painful urination. These symptoms can start even earlier in perimenopause and will not improve after the transitional years unless being explicitly treated. If the symptoms become problematic, they can be cured by applying oestrogen locally, through a cream or a tablet that is introduced in the vagina.

Dream-box

There are two things I want to do when I grow older. The first one is to get up earlier, and the second is to take a swim every morning in spite of the temperature. I have this romantic idea on how wonderful it will be to be up in the early morning hours when the world is still quiet, go out for a swim in the river I live by, and then go home for my morning coffee. My project might already fail on the getting out of bed part. I have, however, started to take swims in colder water recently and, even though I find it terrible to get in, I love the feeling of the blood rush afterwards. There is also anecdotal evidence that cold water swimming can help against perimenopausal symptoms but nothing that has been scientifically proven yet.[666]

Fertility – don't lose control

You are definitely less fertile during perimenopause but as you are still ovulating occasionally, do use contraception, if you do not want to get pregnant.[667] Since using hormonal contraception overrides the hormonal changes, eating them in this phase can even be beneficial as it will eliminate some of the symptoms of perimenopause such as irregular cycles and heavy bleeding. It can even reduce risk of ovarian, endometrial and colorectal cancer.

As a woman over 40, you can safely use hormonal contraception unless you are a smoker, have a BMI above 30, suffer from hypertension or have a family history of cardiovascular disease.[668] The usual risks with hormonal birth control remain the same as before. However, even the lowest contraceptive dose has an oestrogen dose four times greater than the standard postmenopausal dose. With increasing age, the dose-related risks become significant, so once you are menopausal it is important to swap to hormonal menopause treatment instead of staying on the pill.

Onset of menopause – can it be changed?

I always knew that my love for wine would be rewarding one day, I just didn't know what form it would take. It turns out that a frequent consumption of alcohol is associated with a later menopause.[669] If I am representative of the above study population, my menopause might be delayed by a year or two.[670] This is consistent with reports that women who consume alcohol somehow have higher levels of oestrogen, and greater bone density. This is by no means a promotion of alcohol consumption: remember that only small amounts of alcohol have a positive effect, whereas larger ones are harmful. Heavy alcohol consumption can for instance increase the risk of breast cancer.[671] This is a side effect the high levels of oestrogen shares with alcohol.

As always, genetics play a part in the exact timing of menopause but, apart from alcohol, there are a few other lifestyle factors that can have an influence. Smoking is known to lead to an earlier menopause. It increases the FSH level which leads to shorter cycles with, as a consequence, the egg reserve decreasing at a faster pace.[672]

Undernourished women, vegetarians and vegans can also experience an earlier menopause. This could be linked to the contribution of body fat to oestrogen production. More oestrogen inhibits the FSH, so it has the opposite effect of smoking. For the same reason, thinner women might also experience a slightly earlier menopause.[673] So more fat, means more oestrogen, which leads to a later menopause. On the other hand, those with low-fat, plant-based diets have a smaller drop in oestrogen once menopause is reached which can lead to less severe symptoms.

Having more children and breastfeeding for a longer time significantly lowers the risk of early menopause.[674] During pregnancy and breastfeeding, ovulation is temporarily halted so this saves the follicular reserves in your ovaries. In the same way, it would be reasonable to believe that using hormonal contraceptives would delay menopause as it halts ovulation. That is, however, not the case. If you use high-dose contraceptives it can even lead to earlier menopause by as much as three years. As for low-dose contraceptives, they have no influence on the age of menopause.[675] The authors of the study speculate that the older, high-dose contraceptives might have harmed the quality of the follicles in the ovaries, potentially by the type of progestin that was used in the early versions of the pill.

Long term effects of the missing sex hormones: Aging

Aging is not always a walk in the park. The alternative, however, is definitively less appealing. The grey hair you discovered this morning, this new wrinkle by your eye, the shortness of breath: it just means you are alive. And I think that we should be grateful for the long lives we are expected to live nowadays, even if the side effect of long lives is aging. Aging is a very complex mechanism that we are just beginning to understand. It includes many factors linked both to our genes and to our environment. A few of those factors are related to the declining levels of sexual hormones.

As we have previously seen, oestrogen has many positive effects in protecting our bodies. With menopause, the oestrogen suddenly disappears which inevitably has long term consequences. The natural protection that oestrogen and progesterone give us is suddenly gone. The oestrogen was protecting our cardiovascular system against stiff arteries and bad cholesterol, the brain against dementia and Alzheimer's, and our bones against eroding. The oestrogen is responsible for the vascularization of the skin and it stimulates the collagen that makes the skin more elastic. When this effect is gone, the skin starts to age faster and gets drier and more wrinkled.[676]

As we have explained the detailed actions of oestrogen in previous chapters, we will here only go through the consequences of the oestrogen suddenly being removed. If you want a reminder of the detailed actions, feel free to go back to the chapter of the hormonal impact.

Bone: Oestrogen prevents bone loss, and when oestrogen is removed the bone erodes faster than it reconstructs, which is why many older women suffer from osteoporosis. Daily dosing of vitamin D alone does not help the bone but exercise and vitamin D together with calcium may do the job.[677]

Hair: Oestrogen stimulates hair growth by shortening the resting phase and prolonging the growing phase in the hair cycle. Inhibiting oestrogen can therefore lead to hair thinning.[678]

Brain: Oestrogen has a protective effect against Alzheimer's but, once it is gone, women are suddenly much more at risk of getting this disease. As many as 65% of people with Alzheimer's are women. Oestrogen is capable of protecting in many ways. Before menopause, the female brain has a very high glucose metabolism, but it already drops by about 15% during perimenopause. After menopause, it will eventually plunge by 20–30%. This decline in glucose metabolism is due to the oestrogen disappearing. Researchers also believe it is the reason why women develop Alzheimer's at a much higher frequency than men. One woman in five develops the disease, whereas the proportion is only of 1 in 9 for the men.[679] With low oestrogen, the barrier between the brain and the blood is less strong, which can make the brain more susceptible to toxins. There are indications that your lifelong oestrogen exposure can have a link to the risk of Alzheimer's: indeed, the more years with regular menstrual cycles equals the lower the risk of Alzheimer's. However, more research is needed on this topic. For instance, we don't know the impact of using hormonal birth control, a treatment that also alters the overall oestrogen exposure.[680] For women taking oestrogen as HRT after menopause, the risk for Alzheimer's is reduced by 50% to 70%, but only if oestrogen is used before Alzheimer's starts.

Thyroid: Hypothyroidism becomes much more common after menopause.[681] Researchers have shown how oestrogen has a stimulating effect on the thyroid and can also be used to boost the production of thyroid hormone.

Muscles: Even though the cyclic oestrogen changes cannot be linked to muscle strength and muscle mass, the strongly declining levels of oestrogen after menopause do influence the muscles.[682] After menopause, more bone mass is lost than rebuilt, leading to feebler bones. The combination of the weaker bones with weaker muscles is bad, as less muscles also make you more likely to fall, which severely increases your risk of fractures. HRT can prevent this, together with a programme of muscle and balance training.

Heart: Oestrogen used to have a protective effect both on the heart and on the blood vessels, an effect that is now gone. Before menopause, strokes and heart attacks are much more common in men, but after menopause the trend is rapidly reversed.

Urinary tract: Postmenopausal women are especially vulnerable to recurrent urinary tract infections. This can also be linked to the lack of oestrogen since it has been shown that extra oestrogen can help[683] as it can restore the bacterial balance. This is actually a big problem and as many as 55% of postmenopausal women suffer from it.

Many of these changes are rather unfavourable but there is a lot you can do for a healthy aging. Your chronological years are one thing, but your biological age does not have to tick at the same speed. Your lifestyle will have a huge influence on your biological age. As always, it comes down to the same two factors: a lot of movement and a healthy diet. There seem to be a strong consensus among researchers that more than 50% of our aging is linked to lifestyle factors. Walking every day and following a healthy diet can do wonders, and the positive message is it is never too late to start.

HRT-box

Hormonal replacement therapy: With all these problems stemming from the decreased oestrogen levels, it seems natural to want to prevent them by simply adding oestrogen artificially. Such hormone replacement therapy has, however, been through a lot of turmoil over the last decades.

An American study from 2002 – by the Women's Health Initiative (WHI) – set out to study the effects of HRT in postmenopausal women. The first results shocked and worried the public to such an extent that HRT was almost entirely removed from the market. Indeed, the authors of the study have linked HRT with both an increased risk of breast cancer and an increased cardiovascular risk. However, it turned out that many mistakes in the setup of that study were made. First of all, the study focused on women who had been in menopause for a longer time which is not the right target group. The study was a randomized trial, meaning that a group of women got HRT while another got placebo. However, the group with HRT was older than the one with placebo. Two thirds of the women who got HRT were between 60 and 79, whereas the group that should receive HRT was around 50.[684]

The study showed that the risk of breast cancer increased by 26% which sounds dramatic but only means that the risk went from 30 in 10,000 to 38 in 10,000. This demonized HRT. When the data was later reanalysed with groups comprising only women between 50 and 59, they were actually able to show benefits in cancer risk, fracture and basically any risk of death.[685]

The results on the cardiovascular side have also been confusing, and the following studies were not conclusive on whether HRT was beneficial or not. Also, here, this has been mainly due to the timing of the therapy. Since it softens them, oestrogen is good for the arteries. On the other hand, it also tends to form blood clots in the veins. This is why oestrogen therapy is only profitable if it is introduced during the transition to menopause, that is to say before the arteries have become stiff. When they have already stiffened, the addition of oestrogen will increase the risk of blood clots.

HRT mainly consists of two hormones, oestrogen and progesterone. In the WHI report, researchers were suspecting that the progesterone part of the HRT was responsible for the increase of the cardiovascular risk, but they only tested one type of artificial progesterone. As we know from our experience with hormonal birth control, each progestin has a very different effect. Therefore, it is inappropriate to generalize the various effects of progestins.[686]

HRT has a major impact on mood as well but there is some contradiction in the results. The timing of the treatment is here of the utmost importance, since it determines if the impact will be positive or negative.[687] You can start taking HRT as soon as you start experiencing menopausal symptoms.

Although it is definitely not a walk in the park, going through menopause might not be as bad as we sometimes imagine. Many changes in the body will happen but they are all a part of a healthy aging. Especially for women who suffer through their menstrual cycles, it can be quite a liberation. Therefore, the best ending of this chapter might be to cite the menopause monologue of Kristen Scott Thomas in the fabulous BBC series *Fleabag*:

*The menopause comes, and it is the most wonderful f****** thing in the world and, yes, your entire pelvic floor crumbles and you get f****** hot, and no one cares, but then – you're free. No longer a slave, no longer a machine with parts. You're just a person, in business ... It is horrendous but then it's magnificent.*

Some final words

Here ends our hormonal journey and I hope that, like me, you have found it fascinating and were amazed by the discovery of all the transformations the female body undergoes. The body is a marvellous construction, a fine-tuned machine that is able to determine when it is ready for pregnancy and, most astonishingly, even to choose the fittest sperm. So many of the phenomena we find mysterious and inexplicable are in fact direct consequences of the body's hormonal interplay. The changes triggered in the brain have repercussions throughout the entire body – from the brain where it all started influencing our emotional state and wellbeing, to the eyes, the vocal cords, the breasts, the gastrointestinal tract and, of course, the reproductive system. We have had a glimpse on how formidable the body is in its function, and how it adapts and reacts to its environment.

We have learned how the body transforms in puberty to start up the repro-ductive mechanisms and the consequences thereof. We have become aware of how the continuous cyclic transformation of the body to prepare for conception can influence behaviour and mood, reached a better understanding of the functioning of female fertility, and some of the ways this cyclic preparation can become disrupted by external factors. We have seen how a new life can be nourished and brought to completion with the largest transformation a woman's body will ever go through, pregnancy. Hormonal changes can have positive and amazing consequences, but these transition periods are also the most vulnerable. In the last chapter, we found out what happens when the reproductive years come to an end and when we are suddenly deprived of the hormones that have played such a large role since puberty.

Throughout those pages some myths have been debunked whereas in other places stereotypes have been confirmed. Although I have continuously mentioned this, it is worth repeating: most science is based on averages and statistical distributions. This means that even if something is true on average, it does not apply to everyone. Therefore, even though I have revealed truths on women and men, these truths apply only per se to the average woman and the average man. In most cases, such distributions overlap so strongly that if you take one random woman and compare her with one random man, there is still a strong possibility that the roles can be inverted. There is also a lot of room at the edges of those statistical distributions so always remember that 'normal' is a vast concept.

The present and the future

At many places in this book, you have read formulations like 'might lead to', 'could potentially be'. As frustrating as this might be, the reason for this is that many things are still unknown to us. Scientists have started looking at many of those questions and the first hypotheses are often formed but still need more confirmation before they can be considered as 'truths'. There is so much more to explore and to learn about the female body, and research is only getting started.

So, what is the status of the research today? In the *BMJ (former British Medical Journal)* – one of the oldest medical journals and one with the highest impact factor (a measure on how important the journal is considered within its field) – an article from 2020 maps the percentage of articles in that paper that focus on women's health topics. The results are rather discouraging since the occurrence of such papers has only slightly increased over the last 70 years. Some topics in particular are dramatically underrepresented, such as postnatal depression and endometriosis. As you have seen previously in this book, these are indeed important issues that affect a large portion of all women and hopefully we will see more research on this in the future.

Throughout this book it has become obvious that women's bodies are fundamentally different than their male counterparts, and that women are not just smaller men. This is why it is so important to take into account the female aspects in all clinical research – no matter what the topic is. The National Institute of Health (NIH) in the US has published a five-year plan in 2019 to highlight the issues critical to women's health that should be rapidly tackled. It includes understanding the basic differences between females and males and understanding how these differences influence disease symptoms, prevention, and management. It also pertains to the exploration of the impact of such differences on the connection between body and mind and its impact on health and disease. As a part of this program, the NIH will also promote women's careers within science, which is essential not only for obtaining equal opportunities in the workspace but also to advances in science on women's health.

Through the studies of the hormonal axis, we have seen how the environment has a strong impact on us. It is difficult to separate issues linked to our surrounding environment and stress, and what is due to an underlying medical condition. It is therefore important that medical research further investigates the impact of education, socioeconomic status, employment and the social context on women's health. Women's health goes beyond the pure medical aspects and is closely linked to gender equality. According to the World Health Organization (WHO) gender inequality is an important obstacle to everyone's right to health. Realizing the importance of such matters is crucial to change. Gender and equality are an integrated part of WHO sustainable development goals for 2030 which also aim at improving women's health, but it is important to bear in mind how closely linked these factors are.

Many positive developments are already happening. Only recently women's health has started being seen as a lucrative market and a lot of innovation is happening in the field. The innovations cover everything from menstrual health, fertility, sexual health, to breast cancer management. A lot of the innovations are driven by the increased connectivity made possible by mobile apps, a better accessibility to information and the de-stigmatization of women's health topics. Fertility and period tracking apps are collecting huge amounts of data which enables more useful research on women's health issues, such as an increased understanding of fertility and a renewed interest for nonhormonal contraception. At-home fertility tests and hormonal testing is on the rise. Connectivity between apps and wearable devices are driving pregnancy monitoring and foetal health.

Stay curious

Each and every topic I have written about in this book deserves a book on its own. The goal of this one has been to give you an overview, making it easier to connect the dots when you go on reading. My hope is that with these basic facts it will be easier to understand which advice to trust in the future, what makes sense, and what does not.

Information-box

The internet is an incredible resource for finding information on any topic, but unfortunately it is also filled with a lot of myths and misinformation. What to do to make sure you find accurate information? The best way, but in no way the easiest one, is to go back to the original source, i.e., the scientific articles that are published in peer reviewed papers. You can obtain those through PubMed Central, a database from the National Institutes of Health. The problem is that those are often difficult to grasp. They are sometimes also very narrow. If you chose to go the hardcore way, it can therefore be a good idea to focus on review papers from peer reviewed journals as these allow you to see bigger trends and as they often contain valuable summaries of the state of a certain topic.

It is quite demanding, but luckily there are easier ways to obtain reliable information. What I have found very useful are websites from well-renowned institutes such as the Mayo Clinic, Harvard Medical School or the NHS in the UK. Some of the major newspapers, such as the New York Times, also have reliable science sections. Those newspapers have very strict requirements on how they report on scientific advances. I also like some websites very much which you have seen referenced here and there in the book. They contain easily digestible content and despite having fact-checked every little detail I found, I have never discovered anything incorrect on their pages. Such pages also make it easy to fact-check by always linking to the original studies.

One of the things I have appreciated the most during the writing of this book is all the discussions I have had with other women that have so generously opened up to me about their experiences. My most important take-away from all these encounters is that despite all being women and similar in many ways, our experiences are very different. Some women are very reactive to hormonal changes, whereas some seem to barely notice them at all. Some women claim the pill is destroying their mental health and wellbeing, whereas others claim the exact opposite. Some women find pregnancy the most fulfilling experience they ever went through, whereas others hated every minute of it. Some women suffer through perimenopause while being rattled by hormonal fluctuations and hot flashes. Others barely notice the transition and reach menopause with a sense of relief and fulfilment. And no, the difference is not about having a positive attitude or not. The changes go much deeper than that, which I hope this book has made clear. We have to accept we can influence some things while others are out of our control. So, if I can give you one piece of advice, be nice to yourself and don't compare yourself to others. Also remember to be nice to others, who might be having a tougher time than you.

The writing of this book has been an amazing adventure for me. I hope that you have learned important things when you have followed me on this journey, and I hope that you have been inspired to continue your own learning journey.

ACKNOWLEDGEMENTS

For someone used to working in a team, writing a book is solitary work. Since it is your project, and yours alone, asking for help and feedback is difficult. I have been fortunate to have people around me that have been incredibly generous with their time and good will.

First of all, I would like to thank my mum and Julien who have both played such important roles in making this book happen. My mum has helped me with the research and read my texts over and over again. Without her, I might have given up half-way through. Without Julien, it would also have been impossible to pull off this project. His support has been invaluable, both as a never tiring cheerleader and as a head of logistics for the family during this time. Talking of family, my dad also deserves a special thanks for his support and so do my kids.

I would also like to thank my amazing editor Barbara, who did not let me get away with anything. She did not allow me one single sentence that was not fully justified or in any way badly written. The script would not have been the same without her. (She has not edited the acknowledgements so don't hold her responsible for that). A big thank you to Patricia, my designer, that has supported me all the way. When I said I would write the book she immediately offered to do the typesetting and design work. This helped me feel that this was a project that I needed to take all the way and that it would at least look nice in the end.

My friend Anna, the gynaecologist, has read through many of the chapters making sure I am not spreading any misinformation. Thank you so much! My other doctor friends Cristina, Kristina and Tracy have also helped me feel more comfortable with the content by reading other parts of the book. Thank you all!

Thank you to the Birdhaus Writers group with Ana, Elodie, Francesca, Ruzica and Sandra for your continuous support through this process. It has been amazing living this book-writing journey together despite the hardships of life and lockdowns. Thank you also to my early readers: Josefine, Lindsay, Yvonne and Larnie for your valuable feedback and reassurance that this is worth reading.

A special acknowledgement to all of you, family and friends, that chose to share your experiences with me. Without those stories, this book would not be the same, I am so grateful.

Thank you also to Pascal and the Ava-crew for filling me with such courage in the beginning of this process, that I dared to quit my job to write this book. And to everyone else that have supported me on the way, with kind words, encouragements and inspiring curiosity. You are so many that I cannot list you all, but both you and I know who you are. I am very lucky to be surrounded by so many extraordinary people.

References

[1] Bull, J.R., Rowland, S.P., Scherwitzl, E.B. et al. Real-world menstrual cycle characteristics of more than 600,000 menstrual cycles. npj Digit. Med. 2, 83 (2019) doi:10.1038/s41746-019-0152-7

[2] https://www.nejm.org/doi/full/10.1056/NEJMon1211064

[3] Isaacs, E. B., Fischl, B. R., Quinn, B. T., Chong, W. K. Gadian DG. (2010). Impact of breast milk on IQ, brain size and white matter development. Pediatr Res., 67(4), 357–362. https://doi.org/10.1203/PDR.0b013e3181d-026da.Impact

[4] Saini, S. Inferior: How Science Got Women Wrong – and the New Research that's Rewriting the Story. Harper Collins, 2017

[5] Saini, A. Want to do better science? Admit you're not objective. https://www.nature.com/articles/d41586-020-00669-2 [access 2020-11-11]

[6] We've been led to believe that an occasional glass of wine might be better than abstaining from alcohol altogether, but that might not be the case. https://www.bbc.com/future/article/20191021-is-wine-good-for-you?ocid=ww.social.link.whatsapp

[7] We've been led to believe that an occasional glass of wine might be better than abstaining from alcohol altogether, but that might not be the case. https://www.bbc.com/future/article/20191021-is-wine-good-for-you?ocid=ww.social.link.whatsapp

[8] Das, S. K., McIntyre, H. D., Alati, R., Al Mamun, A. Maternal alcohol consumption during pregnancy and its association with offspring renal function at 30 years: Observation from a birth cohort study. Nephrology (Carlton). 2019 Jan;24(1):21-27. doi: 10.1111/nep.13206.

[9] Vitzthum, V. J., Spielvogel, H., Thornburg, J. Interpopulational differences in progesterone levels during conception and implantation in humans. Proc Natl Acad Sci U S A. 2004;101(6):1443-1448. doi:10.1073/pnas.0302640101

[10] Feminizing hormone therapy. https://www.mayoclinic.org/tests-procedures/mtf-hormone-therapy/about/pac-20385096 [access 2020-11-11]

References

[11] Karkazis, K. (2019). The misuses of "biological sex." The Lancet, 394(10212), 1898–1899. https://doi.org/10.1016/S0140-6736(19)32764-3

[12] Criado-Perez C. Invisible Women: Exposing data bias in a world designed for men. Vintage, 2019

[13] Liu, K. A., Mager, N. A. Women's involvement in clinical trials: historical perspective and future implications. Pharm Pract (Granada). 2016;14(1):708. doi:10.18549/PharmPract.2016.01.708

[14] Anderson, G. D, Sex and racial differences in pharmacological response: where is the evidence? Pharmacogenetics, pharmacokinetics, and pharmacodynamics. J Womens Health (Larchmt). 2005 Jan-Feb;14(1):19-29.

[15] Haack, S., Seeringer, A., Thürmann, P.A., Becker, T., Kirchheiner, J. Sex-specific differences in side effects of psychotropic drugs: genes or gender? Pharmacogenomics. 2009 Sep;10(9):1511-26. doi: 10.2217/pgs.09.102.

[16] https://www.fda.gov/drugs/drug-safety-and-availability/questions-and-answers-risk-next-morning-impairment-after-use-insomnia-drugs-fda-requires-lower [access 2020-11-11]

[17] Drug Safety: Most Drugs Withdrawn in Recent Years Had Greater Health Risks for Women GAO-01-286R: Published: Jan 19, 2001. Publicly Released: Feb 9, 2001. [access 2020-11-11.]

[18] http://www.thepipettepen.com/missing-data-how-the-exclusion-of-female-subjects-from-medical-research-hurts-science/ [access 2020-11-11]

[19] https://www.scientificamerican.com/video/the-clitoris-a-reveal-two-millennia-in-the-making/ [access 2020-11-11]

[20] Ramsay, D. T., Kent, J. C, Hartmann, R. A., Hartmann, P. E. Anatomy of the lactating human breast redefined with ultrasound imaging. J Anat. 2005;206(6):525–534. doi:10.1111/j.1469-7580.2005.00417.x

[21] Batrinos, M. L. Testosterone and aggressive behavior in man. Int J Endocrinol Metab. 2012;10(3):563–568. doi:10.5812/ijem.3661

[22] In an interview, Randi Hutter Epstein, a professor at Columbia and author of the book "Aroused: the history of hormones and how they control just about everything", W. W. Norton & Company, used those words to describe how the hormones communicate.

[23] "Hormones". MedlinePlus. U.S. National Library of Medicine. https://medlineplus.gov/hormones.html [access 2020-11-10]

[24] Bryce E., How do your hormones work? Ted-Ed https://www.ted.com/talks/emma_bryce_how_do_your_hormones_work [access 2020-11-10]

[25] Guyton, A. C., Hall, J. E. Textbook of Medical Physiology. Philadelphia, PA: Elsevier Saunders, 11th edition 2006.

[26] Fritz, M. A., Speroff, L. Clinical Gynecologic Endocrinology and Infertility, Philadelphia PA: Lippincott Williams & Wilkins, Eighth edition 2011.

[27] Guyton, A. C., Hall, J. E. (2006). Medical physiology 11th edition. Textbook of Medical Physiology.

[28] "You and your hormones", an education resource from Society of Endocrinology: https://www.yourhormo-

References

nes.info/hormones/prostaglandins/ [access 2020-11-10].

[29] Clark, K., Myatt, L. Prostaglandins and the Reproductive Cycle. The Global Library of Women's Medicine. (ISSN: 1756-2228) 2008; DOI 10.3843/GLOWM.10314

[30] Strassmann, B. (1996). The Evolution of Endometrial Cycles and Menstruation. The Quarterly Review of Biology, 71(2), 181–220.

[31] http://www.bbc.com/earth/story/20150420-why-do-women-have-periods [access 2020-11-12]

[32] Emera, D., Romero, R., Wagner, G. (2012). The evolution of menstruation: A new model for genetic assimilation: Explaining molecular origins of maternal responses to fetal invasiveness. BioEssays, 34(1), 26–35. https://doi.org/10.1002/bies.201100099

[33] Teklenburg, G., Salker, M., Molokhia, M., Lavery, S., Trew, G., Aojanepong, T., et al. (2010). Natural selection of human embryos: Decidualizing endometrial stromal cells serve as sensors of embryo quality upon implantation. PLoS ONE, 5(4), 2–7. https://doi.org/10.1371/journal.pone.0010258

[34] Strassmann, B. (1996). The Evolution of Endometrial Cycles and Menstruation. The Quarterly Review of Biology, 71(2), 181–220.

[35] Diaz, A., Laufer, M. R., & Breech, L. L. (2006). Menstruation in girls and adolescents: Using the menstrual cycle as a vital sign. Pediatrics, 118(5), 2245–2250. https://doi.org/10.1542/peds.2006-2481

[36] Miller, G., Tybur, J. M., & Jordan, B. D. (2007). Ovulatory cycle effects on tip earnings by lap dancers: economic evidence for human estrus? Evolution and Human Behavior, 28(6), 375–381. https://doi.org/10.1016/J.EVOLHUMBEHAV.2007.06.002

[37] Taraborrelli, S. (2015). Physiology, production and action of progesterone. Acta Obstetricia et Gynecologica Scandinavica, 94, 8–16. https://doi.org/10.1111/aogs.12771

[38] Carmina, E., Stanczyk, F. Z., Lobo, R. A. Evaluation of Hormonal Status, Yen & Jaffe's Reproductive Endocrinology (Sixth Edition), 2009. DOI: 10.1016/B978-1-4160-4907-4.00032-2

[39] Guyton, A. C., Hall, J. E. (2006). Medical physiology 11th edition. Textbook of Medical Physiology.

[40] Frobenius W. Ludwig Fraenkel: 'spiritus rector' of the early progesterone research. Eur J Obstet Gynecol Reprod Biol. 1999 Mar;83(1):115-9. doi: 10.1016/s0301-2115(98)00297-8. PMID: 10221621.

[41] Hommel, A., Alexander, H. Zu einigen Aspekten des Lebenswerkes von Ludwig Fraenkel (1870-1951) unter besonderer Berücksichtigung seiner sozialgynäkologischen und sexualwissenschaftlichen Arbeiten [Some aspects of the life work of Ludwig Fraenkel (1870-1951) with special reference to his social gynecology and sexology studies]. Zentralbl Gynakol. 1998;120(10):475-80. German. PMID: 9823647.

[42] Ferin, M. The Hypothalamic-Hypophyseal-Ovarian Axis and the Menstrual Cycle. Glob. libr. women's med.,(ISSN: 1756-2228) 2008; DOI 10.3843/GLOWM.10283 (ISSN: 1756-2228) 2008; DOI 10.3843/GLOWM.10283

[43] Jamil, Z., Fatima, S. S., Ahmed, K., Malik, R. (2016). Anti-Mullerian Hormone: Above and beyond Conventional Ovarian Reserve Markers. Disease Markers, 2016. https://doi.org/10.1155/2016/5246217

References

[44] Ferin, M, Glob. libr. women's med., (ISSN: 1756-2228) 2008; DOI 10.3843/GLOWM.10283

[45] Ghazal, S, Kulp Makarov, J, et al. Egg Transport and Fertilization Glob. libr. women's med., (ISSN: 1756-2228) 2014; DOI 10.3843/GLOWM.10317

[46] Fritz, M.A, Speroff, L. Clinical Gynecologic Endocrinology and Infertility, Eighth edition, Lippincott Williams & Wilkins, 2011

[47] Lobmaier, J. S., Bachofner, L. M. Timing is crucial: Some critical thoughts on using LH tests to determine women's current fertility. Horm Behav. 2018;106:A2-A3. doi:10.1016/j.yhbeh.2018.07.005

[48] Fritz, M.A. and Speroff, L. Clinical Gynecologic Endocrinology and Infertility, Eighth edition, Lippincott Williams & Wilkins, 2011

[49] Ibid.

[50] Kashyap, S, Tran, N, et al. In Vitro Fertilization. Glob. libr. women's med., (ISSN: 1756-2228) 2009; DOI 10.3843/GLOWM.10365

[51] Ibid

[52] Overview of Multiple Pregnancy. https://www.stanfordchildrens.org/en/topic/default?id=overview-of-multiple-pregnancy-85-P08019 [access 2020-11-12]

[53] Gornet, M. E., Lindheim, S.   R., Christianson, M. S. (2019). Ovarian tissue cryopreservation and transplantation: what advances are necessary for this fertility preservation modality to no longer be considered experimental? Fertility and Sterility, 111(3), 473–474. https://doi.org/10.1016/j.fertnstert.2019.01.009

[54] Jensen, A. K., Kristensen, S. G., MacKlon, K. T., Jeppesen, J. V., Fedder, J., Ernst, E., Andersen, C. Y. (2015). Outcomes of transplantations of cryopreserved ovarian tissue to 41 women in Denmark. Human Reproduction, 30(12), 2838–2845. https://doi.org/10.1093/humrep/dev230

[55] New medical procedure could delay menopause by 20 years https://www.theguardian.com/science/2019/aug/04/medical-procedure-delay-menopause [access 2020-11-13]

[56] Ovarian tissue cryopreservation: a committee opinion. Fertility and Sterility, Volume 101, Issue 5, 1237 – 1243

[57] Egg freezing in fertility treatment. Trends and figures: 2010-2016. Human fertilization and embryology authority, UK, Report 2018. https://www.hfea.gov.uk/media/2656/egg-freezing-in-fertility-treatment-trends-and-figures-2010-2016-final.pdf [access 2020-11-13]

[58] https://www.webmd.com/women/guide/mittelschmerz-pain#1

[59] Guyton, A. C., Hall, J. E. (2006). Medical physiology 11th edition. Textbook of Medical Physiology.

[60] Li, S., Winuthayanon, W. (2017). Oviduct: Roles in fertilization and early embryo development. Journal of Endocrinology, 232(1), R1–R26. https://doi.org/10.1530/JOE-16-0302

[61] Druyer et al. Oil-Based or Water-Based Contrast for Hysterosalpingography in Infertile Women. N Engl J Med 2017; 376:2043-2052

[62] Female sterilisation-Your contraception guide. https://www.nhs.uk/conditions/contraception/female-sterili-

References

sation/ [access 2020-11-12]

[63] Fritz, M.A. and Speroff, L. Clinical Gynecologic Endocrinology and Infertility. Eighth edition, Lippincott Williams & Wilkins, 2011 obs ref 19

[64] https://www.wired.com/2014/05/fantastically-wrong-wandering-womb/

[65] Guyton, A. C., Hall, J. E. (2006). Medical physiology 11th edition. Textbook of Medical Physiology. Ref 14

[66] Ibid.

[67] Hendrickson-Jack, L.The fifth vital sign : Master Your Cycles & Optimize Your Fertility Fertility Friday Publishing Inc 2019

[68] Eggert-Kruse, W. (2000). Antimicrobial activity of human cervical mucus. Human Reproduction, 15(4), 778–784. https://doi.org/10.1093/humrep/15.4.778

[69] Weschler, Toni (2006). Taking charge of your fertility : the definitive guide to natural birth control, pregnancy achievement, and reproductive health(Revised ed.). New York, NY: Collins. pp. 59, 64. ISBN 978-0-06-088190-0.

[70] Friedler S, Schenker JG, Herman A, Lewin A. The role of ultrasonography in the evaluation of endometrial receptivity following assisted reproductive treatments: a critical review. Hum Reprod Update 1996; 2: 323–335.

[71] Guyton, A. C., Hall, J. E. (2006). Medical physiology 11th edition. Textbook of Medical Physiology. Ref 14

[72] Alvergne, A., Högqvist Tabor, V. (2018). Is Female Health Cyclical? Evolutionary Perspectives on Menstruation. Trends in Ecology and Evolution, 33(6), 399–414. https://doi.org/10.1016/j.tree.2018.03.006

[73] Cunningham, F. G., Leveno, K. J., Bloom, S. L., Spong, C. Y., Dashe, J. S, Hoffman, B. L., Casey, B. M., Sheffield, J. S. Williams .. 24th ed. New York: McGraw Hill Education; 2014. 84–86 p.

[74] Proctor, M., Farquhar, C. (2006). Diagnosis and management of dysmenorrhoea. British Medical Journal, 332(7550), 1134–1138. https://doi.org/10.1136/bmj.332.7550.1134

[75] Lethaby, A., Augood, C., Duckitt, K. Nonsteroidal anti☐inflammatory drugs for heavy menstrual bleeding. Cochrane Database of Systematic Reviews 1998, Issue 3. Art. No.: CD000400. DOI: 10.1002/14651858.CD000400.

[76] Garry, R., Hart, R., Karthigasu, K. A., Burke, C. (2009). A re-appraisal of the morphological changes within the endometrium during menstruation: A hysteroscopic, histological and scanning electron microscopic study. Human Reproduction, 24(6), 1393–1401. https://doi.org/10.1093/humrep/dep036

[77] Moorman P. G., Myers E. R., et al. (2012). Effect of Hysterectomy With Ovarian Preservation on Ovarian Function. Group, 118(6), 13–14. https://doi.org/10.1097/AOG.0b013e318236fd12.Effect

[78] Endometriosis Symptoms: Neuropathy. https://www.endofound.org/neuropathy [access 2020-11-12]

[79] Kim, A, Adamson, G. Glob. libr. women's med., (ISSN: 1756-2228) 2008; DOI 10.3843/GLOWM.10011

[80] Maccagnano, C., Pellucchi, F., Rocchini, L., Ghezzi, M., Scattoni, V., Montorsi, F., Rigatti, P., Colombo, R.: Diagnosis and Treatment of Bladder Endometriosis: State of the Art. Urol Int 2012;89:249-258. doi: 10.1159/000339519

References

[81] https://www.mayoclinic.org/diseases-conditions/endometriosis/symptoms-causes/syc-20354656 [access 2020-12-12]

[82] What is endometriosis? https://www.fertilityiq.com/endometriosis/what-is-endometriosis [access 2020-12-12]

[83] Zondervan, K. T., Becker, C. M., Missmer, S. A.. (2020). Review Article: Endometriosis. The New England Journal of Medicine, 382, 1244–1256. https://doi.org/10.1177/1461444810365020

[84] Uterine fibroids. https://www.mayoclinic.org/diseases-conditions/uterine-fibroids/symptoms-causes/syc-20354288 [access 2020-11-12]

[85] Fibroids. https://www.healthline.com/health/uterine-fibroids [access 2020-11-12]

[86] Stewart, E. A., Nicholson, W. K., Bradley, L., Borah, B. J. The burden of uterine fibroids for African-American women: results of a national survey. J Womens Health (Larchmt). 2013;22(10):807-816. doi:10.1089/jwh.2013.4334

[87] Adenomyosis. https://www.mayoclinic.org/diseases-conditions/adenomyosis/symptoms-causes/syc-20369138 access 2020-11-12]

[88] Guyton, A. C., Hall, J. E. (2006). Medical physiology 11th edition. Textbook of Medical Physiology.

[89] What Causes Menstrual Clots and Are My Clots Normal? https://www.healthline.com/health/womens-health/menstrual-clots#what's-normal [access 2020-11-12]

[90] Dasharathy, S. S., Mumford, S. L., Pollack, A. Z., Perkins, N. J., Mattison, D. R., Wactawski-Wende, J., & Schisterman, E. F. (2012). Menstrual bleeding patterns among regularly menstruating women. American Journal of Epidemiology, 175(6), 536–545. https://doi.org/10.1093/aje/kwr356

[91] Fraser, I. S., Critchley, H. O. D., Munro, M. G., Broder, M. (2007). Can we achieve international agreement on terminologies and definitions used to describe abnormalities of menstrual bleeding? Human Reproduction, 22(3), 635–643. https://doi.org/10.1093/humrep/del478

[92] Fraser, I. S., Langham, S., Uhl-Hochgraeber, K. (2009). Health-related quality of life and economic burden of abnormal uterine bleeding. Expert Review of Obstetrics and Gynecology, 4(2), 179–189. https://doi.org/10.1586/17474108.4.2.179

[93] Small, C. M., Manatunga, A. K., Klein, M., Feigelson, H. S., Dominguez, C. E., McChesney, R., Marcus, M. (2006). Menstrual Cycle Characteristics. Epidemiology, 17(1), 52–60. https://doi.org/10.1097/01.ede.0000190540.95748.e6

[94] Kolstad, H. A., Bonde, J. P., Hjøllund, N. H., Jensen, T. K., Henriksen, T. B., Ernst, E., et al. (1999). Menstrual cycle pattern and fertility: a prospective follow-up study of pregnancy and early embryonal loss in 295 couples who were planning their first pregnancy. Fertility and Sterility, 71(3), 490–496. https://doi.org/10.1016/S0015-0282(98)00474-9

[95] Jensen, T. K., Scheike, T., Keiding, N., Schaumburg, I., Grandjean, P. (1999). Fecundability in Relation to Body Mass and Menstrual Cycle Patterns.

[96] Fritz, M.A. and Speroff, L. Clinical Gynecologic Endocrinology and Infertility, Eighth edition, Lippincott

Williams & Wilkins, 2011 Obs ref 19

[97] Dasharathy, S. S., Mumford, S. L., Pollack, A. Z., Perkins, N. J., Mattison, D. R., Wactawski-Wende, J., Schisterman, E. F. (2012). Menstrual bleeding patterns among regularly menstruating women. American Journal of Epidemiology, 175(6), 536–545. https://doi.org/10.1093/aje/kwr356

[98] Proctor, M., Farquhar, C. (2006). Diagnosis and management of dysmenorrhoea. British Medical Journal, 332(7550), 1134–1138. https://doi.org/10.1136/bmj.332.7550.1134.

[99] Dawood, M. Y. (2006). Clinical Expert Series Primary Dysmenorrhea Advances in Pathogenesis and Management. Obstet Gynecol, 108(2), 428–441.

[100] Hadfield, R., Mardon, H., Barlow, D., Kennedy, S. (1996) Delay in the diagnosis of endometriosis: a survey of women from the USA and the UK. Hum. Reprod. , 11, 878–880.

[101] Soliman, A. M., Fuldeore, M., Snabes, M., C.. Factors Associated with Time to Endometriosis Diagnosis in the United States. Journal of Women's Health Vol. 26, No. 7 Original Articles

[102] Lethaby, A., Augood, C., Duckitt, K. Nonsteroidal anti☐inflammatory drugs for heavy menstrual bleeding. Cochrane Database of Systematic Reviews 1998, Issue 3. Art. No.: CD000400. DOI: 10.1002/14651858.CD000400.

[103] Wang, W., Knovich, M. A., Coffman, L. G., Torti, F. M., (2010). Serum Ferritin: Past, Present and Future. Biochim Biophys Acta., 1800(8), 760–769. https://doi.org/10.1016/j.cortex.2009.08.003.Predictive

[104] Soppi, E. (2019). Iron Deficiency Without Anemia – Common , Important , Neglected, 5, 1–7. https://doi.org/10.15761/CCRR.1000456

[105] Clancy, K., Nenko, I., Jasienska, G. (2006). Menstruation does not cause anemia: Endometrial thickness correlates positively with erythrocyte count and hemoglobin concentration in premenopausal women American Journal of Human Biology, 18 (5), 710-713 DOI: 10.1002/ajhb.20538

[106] Lowe, R. F., Prata, N. (2013). Hemoglobin and serum ferritin levels in women using copper-releasing or levonorgestrel-releasing intrauterine devices: A systematic review. Contraception, 87(4), 486–496. https://doi.org/10.1016/j.contraception.2012.09.025

[107] Biétry, F. A., Hug, B., Reich, O., Jick, S. S., Meier, C. R. (2017). Iron supplementation in Switzerland – A binational, descriptive, observational study. Swiss Medical Weekly, 147(June). https://doi.org/10.4414/smw.2017.14444

[108] Toxic shock syndrome: B.C. teen's death revives an '80s anxiety. https://www.cbc.ca/news/canada/british-columbia/toxic-shock-syndrome-1.4723453 [access 2020-11-12]

[109] Yang, H., Zhou, B., Prinz, M., Siegel, D. (2012). Proteomic analysis of menstrual blood. Molecular and Cellular Proteomics, 11(10), 1024–1035. https://doi.org/10.1074/mcp.M112.018390

[110] Poliness, A. E., Healy, M. G., Brennecke, S. P., Moses, E. K. Proteomic approaches in endometriosis research. Proteomics. 2004;4:1897–1902.

[111] Warren, L.A., Shih, A., Renteira, S.M. et al. Analysis of menstrual effluent: diagnostic potential for endome-

References

triosis. Mol Med 24, 1 (2018) doi:10.1186/s10020-018-0009-6

[112] Walker, A. E. The Menstrual Cycle. Routledge. London 1997.

[113] Lord, A. M. "The Great Arcana of the Deity": Menstruation and Menstrual Disorders in Eighteenth-Century British Medical Thought." Bulletin of the History of Medicine, vol. 73 no. 1, 1999, p. 38-63. Project MUSE, doi:10.1353/bhm.1999.0036.

[114] The overlooked condition that can trigger extreme behaviour. https://www.bbc.com/future/article/20191213-pmdd-a-little-understood-and-often-misdiagnosed-condition?ocid=global_future_rss [access 2020-11-12]

[115] Fritz, M.A., Speroff, L. Clinical Gynecologic Endocrinology and Infertility, Eighth edition, Lippincott Williams & Wilkins, 2011

[116] The inspiring story of Britain's first female world champion boxer https://www.bbc.com/sport/boxing/52942479 [access 2020-11-12]

[117] Wend, K., Wend, P., Krum, S. A. Tissue-Specific Effects of Loss of Estrogen during Menopause and Aging. Front Endocrinol (Lausanne). 2012;3:19. Published 2012 Feb 8. doi:10.3389/fendo.2012.00019

[118] Hausmann, M., Becker, C., Gather, U., Gunturkun, O. (2002). Functional cerebral asymmetries during the menstrual cycle: a cross-sectional and longi- tudinal analysis. Neuropsychologia 40, 808–816. doi: 10.1016/S0028-3932(01) 00179-8

[119] Hausmann, M. (2017). Why sex hormones matter for neuroscience: A very short review on sex, sex hormones, and functional brain asymmetries. Journal of Neuroscience Research, 95(1–2), 40–49. https://doi.org/10.1002/jnr.23857

[120] How the menstrual cycle changes women's brains – for better. https://www.bbc.com/future/article/20180806-how-the-menstrual-cycle-changes-womens-brains-every-month [access 2020-11-12]

[121] Taraborrelli, S. (2015). Physiology, production and action of progesterone. Acta Obstetricia et Gynecologica Scandinavica, 94, 8–16. https://doi.org/10.1111/aogs.12771

[122] Sayeed, I, Stein, D. Progesterone as a neuroprotective factor in traumatic and ischemic brain injury (2009). Progress in Brain Research. 175, 219-237

[123] Wang, J. M. Regeneration in a degenerating brain: potential of allopregnanolone as a neuroregenerative agent. Curr Alzheimer Res. 2007;4:510-7

[124] Zárate, S., Stevnsner, T., Gredilla, R. Role of Estrogen and Other Sex Hormones in Brain Aging. Neuroprotection and DNA Repair, Frontiers in Aging Neuroscience, 9, 2017,430, https://www.frontiersin.org/article/10.3389/fnagi.2017.00430, DOI=10.3389/fnagi.2017.00430

[125] Wise, P. M., Dubal, D. B.,Wilson M. E., Rau S. W., Böttner, M. Minireview: Neuroprotective Effects of Estrogen—New Insights into Mechanisms of Action, Endocrinology, Volume 142, Issue 3, 1 March 2001, Pages 969–973, https://doi.org/10.1210/endo.142.3.8033

[126] Protopopescu, X., Butler, T., Pan, H., Root, J., Altemus, M., Polaneczky, M., et al. (2008). Hippocampal

structural changes across the menstrual cycle. Hippocampus 18, 985–988. doi: 10.1002/hipo.20468

[127] Pletzer, B., Kronbichler, M., Aichhorn, M., Bergmann, J., Ladurner, G., Kerschbaum, H. H. (2010). Menstrual cycle and hormonal contracep- tive use modulate human brain structure. Brain Res. 1348, 55–62. doi: 10.1016/j.brainres.2010.06.019

[128] De Bondt, T., Van Hecke, W., Veraart, J., Leemans, A., Sijbers, J., Sunaert, S., et al. (2013). Does the use of hormonal contraceptives cause microstructural changes in cerebral white matter? Preliminary results of a DTI and tractography study. Eur. Radiol. 23, 57–64. doi: 10.1007/s00330-012-2572-5

[129] Leeners, B., Kruger, T. H. C., Geraedts, K., Tronci, E., Mancini, T., Ille, F., et al. (2017). Lack of associations between female hormone levels and visuospatial working memory, divided attention and cognitive bias across two consecutive menstrual cycles. Frontiers in Behavioral Neuroscience, 11(July), 1–10. https://doi.org/10.3389/fnbeh.2017.00120

[130] Barth, C., Villringer, A., Sacher, J. (2015). Sex hormones affect neurotransmitters and shape the adult female brain during hormonal transition periods. Frontiers in Neuroscience, 9(FEB), 1–20. https://doi.org/10.3389/fnins.2015.00037

[131] Protopopescu, X., Butler, T., Pan, H., JRoot, J., Altemus M.,et al.. Hippocampal Structural Changes Across the Menstrual Cycle. HIPPOCAMPUS 18:985–988 (2008)

[132] Guéguen, N. (2012). Risk Taking and Women ' s Menstrual Cycle : Near Ovulation , Women Avoid a Doubtful Man, 3(1), 1–3. https://doi.org/10.5178/lebs.2012.17

[133] Iannello, P., Biassoni, F., Nelli, B., Zugno, E., Colombo, B. (2015). The influence of menstrual cycle and impulsivity on risk-taking behavior, (April). https://doi.org/10.7358/neur-2015-017-iann

[134] Stanton, S. J., Mullette-Gillman, O. A., Huettel, S. A. Seasonal variation of salivary testosterone in men, normally cycling women, and women using hormonal contraceptives., Physiol Behav. 2011 Oct 24; 104(5):804-8.

[135] Derntl, B., Pintzinger, N., Kryspin-Exner, I., Schöpf, V. The impact of sex hormone concentrations on decision-making in females and males. Front Neurosci. 2014;8:352. Published 2014 Nov 5. doi:10.3389/fnins.2014.00352

[136] Autism's sex ratio, explained. https://www.spectrumnews.org/news/autisms-sex-ratio-explained/ [access 2020-11-12]

[137] Wetherill, R. R., Jagannathan, K., Hager, N., Maron, M., Franklin, T. R. (2016). Influence of menstrual cycle phase on resting-state functional connectivity in naturally cycling, cigarette-dependent women. Biology of Sex Differences, 7(1), 1–9. https://doi.org/10.1186/s13293-016-0078-6

[138] Metaphor Can Help Patients Understand Receptor Modulation for Antidepressant Medications https://www.psychcongress.com/blog/metaphor-can-help-patients-understand-receptor-modulation-antidepressant-medications pour les sites internet on dit qqch comme (last seen xx.xx.2020) [access 2020-11-12]

[139] Halbreich, U., Backstrom, T., Eriksson, E., O'Brien, S., Calil, H., Ceskova, E., et al. (2007). Clinical diagnostic criteria for premenstrual syndrome and guidelines for their quantification for research studies. Gynecol. Endo-

References

crinol. 23, 123–130. doi: 10.1080/09513590601167969

[140] Saunders, K. E., Hawton, K. (2006). Suicidal behaviour and the menstrual cycle. Psychol. Med. 36, 901–912. doi: 10.1017/S0033291706007392

[141] Nott, P. N., Franklin, M., Armitage, C., Gelder, M. G. (1976). Hormonal changes and mood in the puerperium. Br.J.Psychiatry 128, 379–383. doi: 10.1192/bjp.128.4.379

[142] O'Hara, M. W., Stuart, S., Gorman, L. L., Wenzel, A. (2000). Efficacy of inter- personal psychotherapy for postpartum depression. Arch. Gen. Psychiatry 57, 1039–1045. doi: 10.1001/archpsyc.57.11.1039

[143] Meyer, J. H., Ginovart, N., Boovariwala, A., Sagrati, S., Hussey, D., Garcia, A., et al. (2006). Elevated mono- amine oxidase levels in the brain: an explanation for the monoamine imbalance of major depression. Arch. Gen. Psychiatry 63, 1209–1216. doi: 10.1001/archpsyc.63.11.1209

[144] Barth, C., Villringer, A., Sacher, J. Sex hormones affect neurotransmitters and shape the adult female brain during hormonal transition periods. Front Neurosci. 2015;9:37. Published 2015 Feb 20. doi:10.3389/fnins.2015.00037

[145] Bromberger, J. T., Kravitz, H. M. (2011). Mood and Menopause: Findings from the Study of Women's Health Across the Nation (SWAN) over 10 Years. Obstetrics and Gynecology Clinics of North America, 38(3), 609–625. https://doi.org/10.1016/j.ogc.2011.05.011

[146] Cohen, L. S.,Soares, C. N.,Vitonis,A.F., Otto,M.W., Harlow, B. L. (2006a). Risk for new onset of depression during the menopausal transition: the Harvard study of moods and cycles. Arch.Gen.Psychiatry 63, 385–390. doi: 10.1001/archpsyc.63.4.385

[147] Harlow, B. L., Wise, L. A., Otto, M. W., Soares, C. N., Cohen, L. S. (2003). Depression and its influence on reproductive endocrine and menstrual cycle markers associated with perimenopause: the Harvard Study of Moods and Cycles. Arch. Gen. Psychiatry 60, 29–36. doi: 10.1001/archpsyc.60.1.29

[148] Melrose, S. Seasonal Affective Disorder: An Overview of Assessment and Treatment Approaches. Depress Res Treat. 2015;2015:178564. doi:10.1155/2015/178564

[149] Strasser, B., Gostner, J.M., Fuchs, D. Mood, food, and cognition: role of tryptophan and serotonin. Curr Opin Clin Nutr Metab Care. 2016 Jan;19(1):55-61. doi: 10.1097/MCO.0000000000000237.

[150] Bromberger, J. T., Kravitz, H. M. (2011). Mood and Menopause: Findings from the Study of Women's Health Across the Nation (SWAN) over 10 Years. Obstetrics and Gynecology Clinics of North America, 38(3), 609–625. https://doi.org/10.1016/j.ogc.2011.05.011

[151] Herrera, A. Y., Nielsen, S. E., Mather, M. (2016). Stress-induced increases in progesterone and cortisol in naturally cycling women. Neurobiology of Stress, 3, 96–104. https://doi.org/10.1016/j.ynstr.2016.02.006

[152] Ziomkiewicz, A,, Pawlowski, B., Ellison, P.T., Lipson, S. F., Thune, I., Jasienska, G. Higher luteal progestero- ne is associated with low levels of premenstrual aggressive behavior and fatigue. Biol Psychol. 2012;91:376–82

[153] Gingnell, M., Morell, A., Bannbers, E., Wikström, J., et al..Menstrual cycle effects on amygdala reactivity to

emotional stimulation in premenstrual dysphoric disorder. Horm Behav. 2012 Sep;62(4):400-6. doi: 10.1016/j. yhbeh.2012.07.005. Epub 2012 Jul 17.

[154] Streeter, C. C., Whitfield, T. H., Owen, L., et al. Effects of yoga versus walking on mood, anxiety, and brain GABA levels: a randomized controlled MRS study. J Altern Complement Med. 2010;16(11):1145–1152. doi:10.1089/acm.2010.0007

[155] Backstrom, T., Bixo, M., Johansson, M., Nyberg, S., Ossewaarde, L., Ragagnin, G. et al. (2014). Allopregnanolone and mood disorders. Progress in Neurobiology, 113, 88–94. https://doi.org/10.1016/j.pneurobio.2008.09.009

[156] Ussher, J. M., Perz, J. (2008). Empathy, Egalitarianism and Emotion Work in the Relational Negotiation of PMS: The Experience of Women in Lesbian Relationships. Feminism & Psychology, 18(1), 87–111. https://doi.org/10.1177/0959353507084954

[157] Epperson, C. N., Steiner, M., Hartlage, S. A., Eriksson, E., Schmidt, P. J., Jones, I., et al. (2012a). Premenstrual dysphoric disorder: evidence for a new cate-gory for DSM-5. Am. J. Psychiatry 169, 465–475. doi: 10.1176/appi. ajp.2012.11 081302

[158] Bäckström, T., Andreen, L., Birzniece, V., Björn, I., Johansson, I. M., Nordenstam-Haghjo, M., et al. (2003). The role of hormones and hormonal treatments in premenstrual syndrome. CNS Drugs, 17(5), 325–342. https://doi.org/10.2165/00023210-200317050-00003

[159] Timby, E., Bäckström, T., Nyberg, S., Stenlund, H., Wihlbäck, A-C, Bixo, M. Women with premenstrual dysphoric disorder have altered sensitivity to allopregnanolone over the menstrual cycle compared to controls,Psychopharmacology (2016) 233:2109, 2117, DOI 10.1007/s00213-016-4258-1

[160] Bäckström, T., Türkmen, S., Wahlström, G., Andreen, L. Johansson, I-M. Tolerance to allopregnanolone with focus on the GABA-A receptor. British Journal of Pharmacology, 2011

[161] Backstrom, T., Andreen, L., Birzniece, V., Bjorn, I., Johansson, I. M., Nordenstam- Haghjo, M., et al. (2003). The role of hormones and hormonal treatments in premenstrual syndrome. CNS Drugs 17, 325–342. doi: 10.2165/00023210- 200317050-00003

[162] Yen, J. Y., Tu, H. P., Chen, C. S., Yen, C. F., Long, C. Y., Ko, C. H. (2013). The effect of serotonin 1A receptor polymorphism on the cognitive function of premenstrual dysphoric disorder. Eur. Arch. Psychiatry Clin. Neurosci. 264, 729–739. doi: 10.1007/s00406-013-0466-4

[163] Bäckström, T., Andreen, L., Birzniece, V., Björn, I., Johansson, I. M., Nordenstam-Haghjo, M., et al. (2003). The role of hormones and hormonal treatments in premenstrual syndrome. CNS Drugs, 17(5), 325–342. https://doi.org/10.2165/00023210-200317050-00003

[164] Pluchino, N., Cubeddu, A., Giannini, A., Merlini, S., Cela, V., Angioni, S., Genazzani, A. R. (2009). Progestogens and brain: An update. Maturitas, 62(4), 349–355. https://doi.org/10.1016/j.maturitas.2008.11.023

[165] Kaore, S. N., Langade, D. K., Yadav, V. K., Sharma, P., Thawani, V.R., Sharma, R. Novel actions of progesterone: what we know today and what will be the scenario in the future? J Pharm Pharmacol. 2012;64:1040–62.

References

[166] Shobeiri, F., Araste, F. E., Ebrahimi, R., Jenabi, E., Nazari, M. Effect of calcium on premenstrual syndrome: A double-blind randomized clinical trial. Obstet Gynecol Sci. 2017;60(1):100–105. doi:10.5468/ogs.2017.60.1.100

[167] Parazzini, F., Di Martino, M., Pellegrino, P. (2017). Magnesium in the gynaecological practice: A literature review. Magnesium Research, 30(1), 1–7. https://doi.org/10.1684/mrh.2017.0419

[168] Fathizadeh, N., Ebrahimi, E., Valiani, M., Tavakoli, N., Hojat, M. Evaluating the effect of magnesium and magnesium plus vitamin B6 supplement on the severity of premenstrual syndrome. Iran J Nurs Midwifery Res. 2010 Dec; 15(Suppl1): 401–405.

[169] Purdue-Smithe, A. C., Manson, J. E., Hankinson, S. E., Bertone-Johnson, E. R. (2016). A prospective study of caffeine and coffee intake and premenstrual syndrome. American Journal of Clinical Nutrition, 104(2), 499–507. https://doi.org/10.3945/ajcn.115.127027

[170] McNeil, J., Doucet, É. (2012). Possible factors for altered energy balance across the menstrual cycle: a closer look at the severity of PMS, reward driven behaviors and leptin variations. European Journal of Obstetrics & Gynecology and Reproductive Biology, 163(1), 5–10. https://doi.org/10.1016/j.ejogrb.2012.03.008

[171] Gillings, M. R. (2014). Were there evolutionary advantages to premenstrual syndrome? Evolutionary Applications, 7(8), 897–904. https://doi.org/10.1111/eva.12190

[172] Stute P., Bodmer, C. Ehlert, U., Eltbogen, R., Ging, Streuli, I., von Wolff, M. (2017) Interdisciplinary consensus on management of premenstrual diso A.,rders in Switzerland, Gynecological Endocrinology, 33:5, 342-348, DOI: 10.1080/09513590.2017.1284788

[173] Fritz, M.A. and Speroff, L. Clinical Gynecologic Endocrinology and Infertility,. Eighth edition, Lippincott Williams & Wilkins, 2011

[174] Brandes, J. L. The Influence of Estrogen on Migraine: A Systematic Review. JAMA. 2006;295(15):1824–1830. doi:10.1001/jama.295.15.1824

[175] Penovich, P. E., Helmers, S. Catamenial epilepsy. Int Rev Neurobiol. 2008;83:79-90. doi: 10.1016/S0074-7742(08)00004-4.

[176] Boivin, J., Buntin, L., Kalebic, N., Harrison, C. (2018). What makes people ready to conceive? Findings from the International Fertility Decision-Making Study. Reproductive Biomedicine and Society Online, 6, 90–101. https://doi.org/10.1016/j.rbms.2018.10.012

[177] Bullivant, S. B., Sellergren, S. A., Stern, K., Spencer, N. A., Jacob, S., Mennella, J. A., McClintock, M. K. (2004). Women's sexual experience during the menstrual cycle: Identification of the sexual phase by noninvasive measurement of luteinizing hormone. Journal of Sex Research, 41(1), 82–93. https://doi.org/10.1080/00224490409552216

[178] Roney, J. R., Simmons, Z. L. (2013) Hormonal predictors of sexual motivation in natural menstrual cycles. Hormones and Behavior, 63 (4), 636-645

[179] Haselton, M. G., Gangestad, S. W. Conditional expression of women's desires and men's mate guarding across the ovulatory cycle. Horm Behav. 2006 Apr; 49(4):509-18.

[180] Guéguen, N. (2008). The receptivity of women to courtship solicitation across the menstrual cycle: a field experiment Biological psychology 80(3):321-4.

[181] Ziegler, T. E. (2007). "Female sexual motivation during non-fertile periods: a primate phenomenon". Hormones and Behavior. 51 (1): 1–2.

[182] Jones, B. C., Little, A. C., Boothroyd, L., Debruine, L.M., Feinberg, D. R., Smith, M. J., Cornwell, R. E., Moore, F. R., Perrett, D. I. (2005). Commitment to relationships and preferences for femininity and apparent health in faces are strongest on days of the menstrual cycle when progesterone level is high. Horm Behav. Sep; 48(3):283-90.

[183] Hiller, J. (2017). Gender differences in sexual motivation. Journal of Men's Health & Gender(Vol. 2, Issue 3)

[184] Mitrokostas, S. (2019). Here's What Happens to Your Body And Brain When You Orgasm. Business Insider, January. https://www.sciencealert.com/here-s-what-happens-to-your-brain-when-you-orgasm. [access 2020-11-12]. [access 2020-11-12]

[185] Battaglia, C., Nappi, R. E., Mancini, F., Cianciosi, A., Persico, N., Busacchi, P., Facchinetti, F., de Aloysio, D. Menstrual cycle-related morphometric and vascular modifications of the clitoris. J Sex Med. 2008 Dec;5(12):2853-61. doi: 10.1111/j.1743-6109.2008.00972.x. Epub 2008 Aug 28. PMID: 18761595.

[186] Baker, R. R., Bellis, M. A. Human sperm competition: ejaculate manipulation by females and a function for the female orgasm. Animal Behaviour Volume 46, Issue 5, November 1993, Pages 887-909

[187] https://www.ted.com/talks/mary_roach_10_things_you_didn_t_know_about_orgasm. [access 2030-11-12]

[188] Buisson, O., Jannini, E. A. (2013). Pilot echographic study of the differences in clitoral involvement following clitoral vaginal sexual stimulation. J Sex Med. 2013 Nov;10(11):2734-40. doi: 10.1111/jsm.12279. Epub 2013 Aug 12.

[189] Pavlicev, M., Wagner, G., 2016. The evolutionary origin of femaleorgasm. J. Exp. Zool. (Mol. Dev. Evol.) 326B:326–337. doi: 10.1002/jez.b.22690

[190] Larson, C. M., Haselton, M. G., Gildersleeve, K. A., Pillsworth, E. G. (2013). Changes in women's feelings about their romantic relationships across the ovulatory cycle. Hormones and Behavior, 63(1), 128–135. https://doi.org/10.1016/j.yhbeh.2012.10.005

[191] Haselton, M. Hormonal. How Hormones drive desire, shape relationships, and make us wiser. 2018, Oneworld publications.

[192] Brown, S. G., Morrison, L. A., Calibuso, M. J., Christiansen, T. M. (2008). The menstrual cycle and sexual behavior: Relationship to eating, exercise, sleep, and health patterns. Women and Health, 48(4), 429–444. https://doi.org/10.1080/03630240802575179

[193] Guyton, A. C., Hall, J. E. Textbook of Medical Physiology. Philadelphia, PA: Elsevier Saunders, 11th edition 2006.

[194] McNeil, J., Doucet, É. (2012). Possible factors for altered energy balance across the menstrual cycle: a closer

References

look at the severity of PMS, reward driven behaviors and leptin variations. European Journal of Obstetrics & Gynecology and Reproductive Biology, 163(1), 5–10. https://doi.org/10.1016/j.ejogrb.2012.03.008

[195] Leeners, B., Geary, N., Tobler, P. N., Asarian, L. (2017). Ovarian hormones and obesity. Human Reproduction Update, 23(3), 300–321. https://doi.org/10.1093/humupd/dmw045

[196] Muccioli, M. Insulin requirements throughout the menstrual cycle, https://www.diabetesdaily.com/learn-about-diabetes/treatment/insulin-101/how-to-use-insulin/insulin-requirements-throughout-the-menstrual-cycle/ [access 2020-12-06]

[197] Leptin. https://www.yourhormones.info/hormones/leptin/ [access 2020-12-07]

[198] Davidsen, L., Vistisen, B., Astrup, A. (2007). Impact of the menstrual cycle on determinants of energy balance: A putative role in weight loss attempts. International Journal of Obesity, 31(12), 1777–1785. https://doi.org/10.1038/sj.ijo.0803699

[199] Ahrens, K., Mumford, S. L., Schliep, K. C., Kissell, K. A., Perkins, N. J., Wactawski-Wende, J., Schisterman, E. F. (2014). Serum leptin levels and reproductive function during the menstrual cycle. American Journal of Obstetrics and Gynecology, 210(3), 248.e1-248.e9. https://doi.org/10.1016/j.ajog.2013.11.009

[200] How drugs affect neurotransmitters. https://thebrain.mcgill.ca/flash/i/i_03/i_03_m/i_03_m_par/i_03_m_par_alcool.html#drogues [access 2020-12-07]

[201] Discovering the Sweet Mysteries of Chocolate. http://faculty.washington.edu/chudler/choco.html [access 2020-12-07]

[202] Leeners, B., Geary, N., Tobler, P. N., Asarian, L. (2017). Ovarian hormones and obesity. Human Reproduction Update, 23(3), 300–321. https://doi.org/10.1093/humupd/dmw045

[203] McNeil, J., Doucet, É. (2012). Possible factors for altered energy balance across the menstrual cycle: a closer look at the severity of PMS, reward driven behaviors and leptin variations. European Journal of Obstetrics & Gynecology and Reproductive Biology, 163(1), 5–10. https://doi.org/10.1016/j.ejogrb.2012.03.008

[204] Davidsen, L., Vistisen, B., Astrup, A. (2007). Impact of the menstrual cycle on determinants of energy balance: A putative role in weight loss attempts. International Journal of Obesity, 31(12), 1777–1785. https://doi.org/10.1038/sj.ijo.0803699

[205] Moore J, Barlow D, Jewell D, Kennedy S. Do gastrointestinal symptoms vary with menstrual cycle? Br J Obstet Gynaecol 1998; 105: 1322–5

[206] Bernstein, M. T., Graff, L. A., Avery, L., Palatnick, C., Parnerowski, K., Targownik, L. E. . (2014). Gastrointestinal symptoms before and during menses in women with IBD. BMC Women's Health volume 14, Article number: 14 . Retrieved from http://www.biomedcentral.com/1472-6874/14/14

[207] Canavan, C., West, J., Card, T. The epidemiology of irritable bowel syndrome. Clin Epidemiol. 2014;6:71–80. Published 2014 Feb 4. doi:10.2147/CLEP.S40245

[208] Pascal, M., Perez-Gordo, M., Caballero, T., et al. Microbiome and Allergic Diseases. Front Immunol.

2018;9:1584. Published 2018 Jul 17. doi:10.3389/fimmu.2018.01584

[209] Menon, R., Watson, S. E., Thomas, L. N. et al. Diet complexity and estrogen receptor β status affect the composition of the murine intestinal microbiota. Appl Environ Microbiol 2013; 79: 5763–73.

[210] rezza, M., di Padova, C., Pozzato, G., Terpin, M., Baraona, E., Lieber, C. S. High Blood Alcohol Levels in Women — The Role of Decreased Gastric Alcohol Dehydrogenase Activity and First-Pass Metabolism, January 11, 1990

N Engl J Med 1990; 322:95-99, DOI: 10.1056/NEJM199001113220205

[211] Evans, S. M., Levin, F.R. Response to alcohol in women: role of the menstrual cycle and a family history of alcoholism. Drug Alcohol Depend. 2011;114(1):18–30. doi:10.1016/j.drugalcdep.2010.09.001

[212] Rachdaoui, N., Sarkar, D. K. Effects of alcohol on the endocrine system. Endocrinol Metab Clin North Am. 2013;42(3):593–615. doi:10.1016/j.ecl.2013.05.008

[213] Travis, R. C., Key, T. J. Oestrogen exposure and breast cancer risk. Breast Cancer Res. 2003;5(5):239–247. doi:10.1186/bcr628

[214] Kinney, A. Kline, J. Kelly, A., Reuss, M. L., Levin, B. Smoking, alcohol and caffeine in relation to ovarian age during the reproductive years, Human Reproduction, Volume 22, Issue 4, April 2007, Pages 1175–1185, https://doi.org/10.1093/humrep/del496

[215] The 'Perfect' Human Body Is Not What You Think. https://www.livescience.com/62895-building-the-perfect-body.html [access 2020-12-06]

[216] Guyton, A. C. Hall, J. E. Textbook of Medical Physiology. Philadelphia, PA: Elsevier Saunders, 11th edition 2006.

[217] Wells, J. C. K. (2012). The evolution of human adiposity and obesity: Where did it all go wrong? DMM Disease Models and Mechanisms, 5(5), 595–607. https://doi.org/10.1242/dmm.009613

[218] Function. https://www.medicalnewstoday.com/articles/277177#function [access 2020-12-07]

[219] Jasieńska, G., Ziomkiewicz, A., Ellison, P. T., Lipson, S. F., Thune, I. (2004). Large breasts and narrow waists indicate high reproductive potential in women. Proceedings of the Royal Society B: Biological Sciences, 271(1545), 1213–1217. https://doi.org/10.1098/rspb.2004.2712

[220] Grillot, R. L., Simmons, Z. L., Lukaszewski, A. W., Roney, J. R. Hormonal and morphological predictors of women's body attractiveness https://doi.org/10.1016/j.evolhumbehav.2014.01.001

[221] ASRM Practice Committee report (2015). Obesity and reproduction: A committee opinion. Fertility and Sterility, 104(5), 1116–1126. https://doi.org/10.1016/j.fertnstert.2015.08.018

[222] Santen, R. J., Benign Breast Disease in Women, www.endotext.org

[223] Eren, T., Aslan, A., Ozemir, I. A., Baysal, H., Sagiroglu, J., Ekinci, O., Alimoglu, O. (2016). Factors effecting mastalgia. Breast Care, 11(3), 188–193. https://doi.org/10.1159/000444359

[224] Santen, R. J. Benign Breast Disease in Women, www.endotext.org

References

[225] Eren, T., Aslan, A., Ozemir, I. A., Baysal, H., Sagiroglu, J., Ekinci, O., Alimoglu, O. (2016). Factors effecting mastalgia. Breast Care, 11(3), 188–193. https://doi.org/10.1159/000444359

[226] Kanis, J. A., Johnell, O., Oden, A., Sembo, I., Redlund-Johnell, I., Dawson, A. et al. Longterm risk of osteoporotic fracture in Malmo. Osteoporos Int 2000;11(8):669-74.

[227] Riggs, B.L., Khosla, S., Melton, L. J. III. Sex steroids and the construction and conservation of the adult skeleton. Endocrin Rev 2002; 23: 279-302.

[228] Lerner, U. H. Bone remodeling in post-menopausal osteoporosis. J Dent Res. 2006; 85:584-95.

[229] Riggs, B.L., Khosla S., Melton, L. J III. Sex steroids and the construction and conservation of the adult skeleton. Endocrin Rev 2002; 23: 279-302.

[230] Shufelt, C. L. Torbati, T., Dutra, E. (2017). Hypothalamic Amenorrhea and the Long-Term Health Consequences. Semin Reprod Med., 35(3), 256–262. https://doi.org/10.1016/j.physbeh.2017.03.040

[231] Smith, S. M., Abrams, S. A., Davis-Street, J. E., Heer, M., O'Brien, K. O., Wastney, M. E., Zwart, S. R. Fifty years of human space travel: implications for bone and calcium research. Annu. Rev. Nutr. 2014. 34:377–400

[232] Iwamoto, J., Takeda1, T., Sato, Y. Prevention of bone loss during space flight. Keio J Med. 2005, 54 (2): 55–59.

[233] Tam, A., Morrish, D., Wadsworth, S., Dorscheid, D., Man, S. F., Sin, D. D. The role of female hormones on lung function in chronic lung diseases. BMC Womens Health. 2011;11:24. Published 2011 Jun 3. doi:10.1186/1472-6874-11-24

[234] Haggerty, C. L., Ness, R. B., Kelsey, S., Waterer, G. W. The impact of estrogen and progesterone on asthma. Ann Allergy Asthma Immunol. 2003;90:284–91. doi: 10.1016/S1081-1206(10)61794-2. quiz 291-3, 347.

[235] Fritz, M. A., Speroff, L. Clinical Gynecologic Endocrinology and Infertility, Philadelphia PA: Lippincott Williams & Wilkins, Eighth edition 2011.

[236] Liu, K. A., Mager, N. A. Women's involvement in clinical trials: historical perspective and future implications. Pharm Pract (Granada). 2016;14(1):708. doi:10.18549/PharmPract.2016.01.708

[237] Tam, A., Morrish, D., Wadsworth, S., Dorscheid, D., Man, S. F., Sin, D. D. The role of female hormones on lung function in chronic lung diseases. BMC Womens Health. 2011;11:24. Published 2011 Jun 3. doi:10.1186/1472-6874-11-24

[238] American Heart Association. Warning signs of a heart attack. https://www.heart.org/en/health-topics/heart-attack/warning-signs-of-a-heart-attack [access 2020-12-07]

[239] Heron, M. Deaths: Leading causes for 2016. National Vital Statistics Reports. 2018;67(6):1-77 .

[240] Temple, N. J. Fat, Sugar, Whole Grains and Heart Disease: 50 Years of Confusion. Nutrients. 2018; 4;10.

[241] Greenwood, B., Carnahan, S., Huang, L. Patient–physician gender concordance and increased mortality among female heart attack patients. PNAS. 2018;115(34):8569-8574. doi: 10.1073/pnas.1800097115.

[242] Iorga, A., Cunningham, C.M., Moazeni, S. et al. Biol Sex Differ (2017) 8: 33. https://doi.org/10.1186/s13293-017-0152-8

[243] Arnold, A. P., Cassis, L. A., Maghbali, M., Reue, K., Sandberg, K. Sex Hormones and Sex Chromosomes Cause Sex Differences in the Development of Cardiovascular Diseases. Arterioscler Thromb Vasc Biol. 2017;37:746-756. DOI: 10.1161/ATVBAHA.116.307301.

[244] Walli-Attaei, M., Joseph, P., Rosengren, A., Chow, C. K., Rangarajan, S., Lear, S. A., et al. Variations between women and men in risk factors, treatments, cardiovascular disease incidence, and death in 27 high-income, middle-income, and low-income countries (PURE): a prospective cohort study https://www.thelancet.com/pdfs/journals/lancet/PIIS0140-6736(20)30543-2.pdf [access 2020-12-07]

[245] Marsh, S. a, Jenkins, D. G. (2002). Physiological responses to the menstrual cycle: implications for the development of heat illness in female athletes. Sports Medicine, 32(10), 601–614.

[246] How period tracking can give all female athletes an edge https://www.theguardian.com/lifeandstyle/shortcuts/2019/jul/10/how-period-tracking-can-give-all-female-athletes-an-edge [access 2020-12-07]

[247] Balachandar, V., Marciniak, J. L., Wall, O., Balachandar, C. Effects of the menstrual cycle on lower-limb biomechanics, neuromuscular control, and anterior cruciate ligament injury risk: a systematic review. Muscles Ligaments Tendons J. 2017;7(1):136–146. Published 2017 May 10. doi:10.11138/mltj/2017.7.1.136

[248] Lei, T. H., Stannard, S. R., Perry, B. G., Schlader, Z. J., Cotter, J. D., Mündel, T. (2017). Influence of menstrual phase and arid vs. humid heat stress on autonomic and behavioural thermoregulation during exercise in trained but unacclimated women. Journal of Physiology, 595(9), 2823–2837. https://doi.org/10.1113/JP273176

[249] Wikström-Frisén, L., Boraxbekk, C. J., Henriksson-Larsén, K. (2017). Effects on power, strength and lean body mass of menstrual/oral contraceptive cycle based resistance training. Journal of Sports Medicine and Physical Fitness, 57(1–2), 43–52. https://doi.org/10.23736/S0022-4707.16.05848-5

[250] Lowe, D. A., Baltgalvis, K. A., Greising, S. M. (2010). Mechanisms behind Estrogens' Beneficial Effect on Muscle Strength in Females. Exerc Sport Sci Rev, 38(2), 61–67. https://doi.org/10.1097/JES.0b013e3181d496bc. Mechanisms

[251] Stening, K., Eriksson, O., Wahren, L. K., Berg, G., Hammar, M., Blomqvist, A. (2007). Pain sensations to the cold pressor test in normally menstruating women: Comparison with men and relation to menstrual phase and serum sex steroid levels. American Journal of Physiology – Regulatory Integrative and Comparative Physiology, 293(4), 1711–1716. https://doi.org/10.1152/ajpregu.00127.2007

[252] Aloisi, A. M. Gonadal hormones and sex differences in pain reactivity. Clin J Pain 2003; 19: 168–74

[253] Goodale, B. M., Shilaih, M., Falco, L., Dammeier, F., Hamvas, G., Leeners, B. (2019). Wearable sensors reveal menses-driven changes in physiology and enable prediction of the fertile window: an observational study. Journal of Medical Internet Research, 21(4), e13404.

[254] Stachenfeld, N. S., Silva, C., Keefe, D. L. (2000). Estrogen modifies the temperature effects of progesterone. Journal of Applied Physiology, 88(5), 1643–1649.

[255] Fernandes, V. S., Ribeiro, A. S., Martínez-Sáenz, A., Blaha, I., Serrano-Margüello, D., Recio, P., Martínez, A. C., t al. Underlying mechanisms involved in progesterone-induced relaxation to the pig bladder neck. Eur J

References

Pharmacol. 2014 Jan 15;723:246-52. doi: 10.1016/j.ejphar.2013.11.025...

[256] Elliott-Sale, K.J., McNulty, K.L., Ansdell, P. et al. The Effects of Oral Contraceptives on Exercise Performance in Women: A Systematic Review and Meta-analysis. Sports Med 50, 1785–1812 (2020). https://doi.org/10.1007/s40279-020-01317-5

[257] Gerhardt, U., Hillebrand, U., Mehrens, T., Hohage, H. (2000). Impact of estradiol blood concentrations on skin capillary Laser Doppler flow in premenopausal women. International Journal of Cardiology, 75(1), 59–64. Retrieved from http://cat.inist.fr/?aModele=afficheN&cpsidt=788772

[258] Guyton, A. C., Hall, J. E. Textbook of Medical Physiology. Philadelphia, PA: Elsevier Saunders, 11th edition 2006.

[259] Ashcroft, G. S., Ashworth, J. J. Potential role of estrogens in wound healing. Am J Clin Dermatol. 2003; 4(11):737-43

[260] Glowing Skin During Pregnancy: Why It Happens https://www.healthline.com/health/pregnancy-glow [access 2020-12-07]

[261] Geller, L., Rosen, J., Frankel, A., Goldenberg, G. (2014). Perimenstrual flare of adult acne. Journal of Clinical and Aesthetic Dermatology, 7(8), 30–34.

[262] Raychaudhuri, S. P., Navare, T., Gross, J., Raychaudhuri, S. K.Clinical course of psoriasis during pregnancy. Int J Dermatol. 2003 Jul; 42(7):518-20

[263] Wend, K., Wend, P., Krum, S. A. Tissue-Specific Effects of Loss of Estrogen during Menopause and Aging. Front Endocrinol (Lausanne). 2012;3:19. Published 2012 Feb 8. doi:10.3389/fendo.2012.00019

[264] Raghunath, R. S., Venables, Z. C., Millington, G. W. M. (2015). The menstrual cycle and the skin. Clinical and Experimental Dermatology, 40(2), 111–115. https://doi.org/10.1111/ced.12588

[265] Stevenson, S., Thornton, J. (2007). Effect of estrogens on skin aging and the potential role of SERMs. Clinical Interventions in Aging, 2(3), 283–297. https://doi.org/10.1080/13697130701467157

[266] Skin and the cycle: how hormones affect your skin https://helloclue.com/articles/cycle-a-z/skin-and-the-cycle-how-hormones-affect-your-skin [access 2020-12-07]

[267] Hutchinson, C. V., Walker, J. A., Davidson, C. (2014). Oestrogen, ocular function and low-level vision: A review. Journal of Endocrinology, 223(2), R9–R18. https://doi.org/10.1530/JOE-14-0349

[268] Webb, A. L. M., Hibbard, P. B., O'Gorman, R. (2018). Natural variation in female reproductive hormones does not affect contrast sensitivity. Royal Society Open Science, 5(2). https://doi.org/10.1098/rsos.171566

[269] Çelik, Ö.,, Çelik, A., Ateşpare, A., Boyacı, Z., Çelebi, S., Gündüz, T., Aksungar, F. B., Yelken, K. Voice and speech changes in various phases of menstrual cycle, J Voice. 2013 Sep;27(5):622-6. doi: 10.1016/j.jvoice.2013.02.006. Epub 2013 Mar 15.

[270] Banai, P. I. (2017). Voice in different phases of menstrual cycle among naturally cycling women and users of hormonal contraceptives. PLoS ONE, 12(8), 1–13. https://doi.org/10.1371/journal.pone.0183462

[271] Pisanski, K., Bhardwaj, K., Reby, D. (2018). Women's voice pitch lowers after pregnancy. Evolution and Human Behavior, 39(4), 457–463. https://doi.org/10.1016/j.evolhumbehav.2018.04.002

[272] https://www.nytimes.com/2020/02/20/health/coronavirus-men-women.html

[273] https://www.washingtonpost.com/climate-environment/2020/03/19/coronavirus-kills-more-men-than-women/

[274] The surprising reason you feel awful when you're sick – Marco A. Sotomayor. TedEd video. https://ed.ted.com/lessons/the-surprising-reason-you-feel-awful-when-you-re-sick-marco-a-sotomayor

[275] Influenza A virus subtype H1N1

[276] Klein, S. L. (2012). Immune cells have sex and so should journal articles. Endocrinology, 153(6), 2544–2550. https://doi.org/10.1210/en.2011-2120

[277] Schurz, H., Salie, M., Tromp, G. et al. The X chromosome and sex-specific effects in infectious disease susceptibility. Hum Genomics 13, 2 (2019). https://doi.org/10.1186/s40246-018-0185-z

[278] Taneja V. Sex Hormones Determine Immune Response. Front Immunol. 2018;9:1931. Published 2018 Aug 27. doi:10.3389/fimmu.2018.01931

[279] Can castration really prolong a man's life? https://www.newscientist.com/article/dn22302-can-castration-really-prolong-a-mans-life/ [access 2020-12-07]

[280] Hughes, G. C. (2012). Progesterone and autoimmune disease. Autoimmunity Reviews, 11(6–7), A502–A514. https://doi.org/10.1016/j.autrev.2011.12.003

[281] Vasiadi, M., Kempuraj, D., Boucher, W., Kalogeromitros, D., Theoharides, T. C. Progesterone inhibits mast cell secretion. Int J Immunopathol Pharmacol 2006; 19: 787–94

[282] Kirmaz, C., Yuksel, H., Mete, N., Bayrak, P., Baytur, Y. B. Is the menstrual cycle affecting the skin prick test reactivity?, Asian Pac J Allergy Immunol. 2004 Dec;22(4):197-203.

[283] Klein, S. L. (2012). Immune cells have sex and so should journal articles. Endocrinology, 153(6), 2544–2550. https://doi.org/10.1210/en.2011-2120

[284] Billington, W. D. The immunological problem of pregnancy: 50 years with the hope of progress. A tribute to Peter Medawar. J Reprod Immunol 2003;60: 1–11

[285] Fleischman, D. S., Fessler, D. M. T. (2011). Progesterone's effects on the psychology of disease avoidance: Support for the compensatory behavioral prophylaxis hypothesis. Hormones and Behavior, 59(2), 271–275. https://doi.org/10.1016/j.yhbeh.2010.11.014

[286] Klein, S. L. (2012). Immune cells have sex and so should journal articles. Endocrinology, 153(6), 2544–2550. https://doi.org/10.1210/en.2011-2120

[287] Khan, D. Ahmed, S. A. (2016) The immune system is a natural target for estrogen action: opposing effects of estrogen in two prototypical autoimmune diseases. Front. Immunol. 6, 635

[288] Klein, S. L. (2012). Immune cells have sex and so should journal articles. Endocrinology, 153(6), 2544–2550.

References

https://doi.org/10.1210/en.2011-2120

[289] Bradley, F., Birse, K., Hasselrot, K., Noël-Romas, L., Introini, A., Wefer, H., et al. (2018). The vaginal microbiome amplifies sex hormone-associated cyclic changes in cervicovaginal inflammation and epithelial barrier disruption. American Journal of Reproductive Immunology, 80(1), 1–13. https://doi.org/10.1111/aji.12863

[290] Bradley, F., Birse, K., Hasselrot, K., Noël-Romas, L., Introini, A., Wefer, H., et al. (2018). The vaginal microbiome amplifies sex hormone-associated cyclic changes in cervicovaginal inflammation and epithelial barrier disruption. American Journal of Reproductive Immunology, 80(1), 1–13. https://doi.org/10.1111/aji.12863

[291] Hel, Z., Stringer, E., Mestecky, J. Sex steroid hormones, hormonal contraception, and the immunobiology of human immunodeficiency virus-1 infection. Endocr Rev 2010;31:79–97.

[292] Khan, D., Ansar Ahmed, S. (2016). The immune system is a natural target for estrogen action: Opposing effects of estrogen in two prototypical autoimmune diseases. Frontiers in Immunology, 6(JAN), 1–8. https://doi.org/10.3389/fimmu.2015.00635

[293] Hughes, G. C. (2012). Progesterone and autoimmune disease. Autoimmunity Reviews, 11(6–7), A502–A514. https://doi.org/10.1016/j.autrev.2011.12.003

[294] Robinson, D. P., Klein, S. L. "Pregnancy and pregnancy-associated hormones alter immune responses and disease pathogenesis." Hormones and behavior vol. 62,3 (2012): 263-71. doi:10.1016/j.yhbeh.2012.02.023

[295] Hughes, G. C. (2012). Progesterone and autoimmune disease. Autoimmunity Reviews, 11(6–7), A502–A514. https://doi.org/10.1016/j.autrev.2011.12.003

[296] Klein, S. L. (2012). Immune cells have sex and so should journal articles. Endocrinology, 153(6), 2544–2550. https://doi.org/10.1210/en.2011-2120

[297] Risk Factors for Cancer. https://www.cancer.gov/about-cancer/causes-prevention/risk [access 2020-12-06]

[298] Jolie, A., Pitt, J.: Diary of a Surgery, The New York Times, March 24, 2015

[299] Travis, R. C., Key, T. J. Oestrogen exposure and breast cancer risk. Breast Cancer Res. 2003;5(5):239–247. doi:10.1186/bcr628

[300] Ibid.

[301] Ibid.

[302] How chronic inflammation can lead to cancer. http://news.mit.edu/2015/how-chronic-inflammation-can-lead-to-cancer-0807 [access 2020-12-07]

[303] Collaborative Group on Epidemiological Studies on Endometrial Cancer. Endometrial cancer and oral contraceptives: an individual participant meta-analysis of 27 276 women with endometrial cancer from 36 epidemiological studies. Lancet Oncol. 2015 Sep;16(9):1061-1070. doi: 10.1016/S1470-2045(15)00212-0. Epub 2015 Aug 4.

[304] Iversen, L., Fielding, S., Lidegaard, Ø., Mørch, L. S., Skovlund, C. W., Hannaford, P. C. et al. Association between contemporary hormonal contraception and ovarian cancer in women of reproductive age in Denmark: prospective, nationwide cohort study BMJ 2018; 362 :k3609

[305] Alvergne, A., Högqvist Tabor, V. (2018). Is Female Health Cyclical? Evolutionary Perspectives on Menstruation. Trends in Ecology and Evolution, 33(6), 399–414. https://doi.org/10.1016/j.tree.2018.03.006

[306] Stanton, S. J., Mullette-Gillman, O. A., Huettel, S. A. Seasonal variation of salivary testosterone in men, normally cycling women, and women using hormonal contraceptives. Physiol Behav. 2011 Oct 24; 104(5):804-8.

[307] Derntl, B., Pintzinger, N., Kryspin-Exner, I., Schöpf, V. The impact of sex hormone concentrations on decision-making in females and males. Front Neurosci. 2014;8:352. Published 2014 Nov 5. doi:10.3389/fnins.2014.00352

[308] Zumoff, B., Strain, G. W., Miller, L. K., et al. Twenty-four-hour mean plasma testosterone concentration declines with age in normal premenopausal women. J Clin Endocrinol Metab.. 1995;80:1429-1430

[309] Nieschlag, E. The history of testosterone, Endocrine Abstracts (2005) 10 S2

[310] Guyton, A. C., Hall, J. E. Textbook of Medical Physiology. Philadelphia, PA: Elsevier Saunders, 11th edition 2006.

[311] Bancroft, J.The endocrinology of sexual arousal. Journal of Endocrinology 186 (3) 411-427 DOI: https://doi.org/10.1677/joe.1.06233

[312] Mushayandebvu, D. V., Castracane, T., Gimpel, T., et al. Evidence for diminished midcycle ovarian androgen production in older reproductive aged women. J Fertil Steril.. 1991;65:721-723.

[313] Bancroft, J.The endocrinology of sexual arousal. Journal of Endocrinology 186 (3) 411-427 DOI: https://doi.org/10.1677/joe.1.06233

[314] Wilcox, A. Fertility and pregnancy. An epidemiologic perspective. Oxford University Press, 2010.

[315] Ibid.

[316] Wu, H., Marwah, S., Wang, P. et al. Misoprostol for medical treatment of missed abortion: a systematic review and network meta-analysis. Sci Rep 7, 1664 (2017). https://doi.org/10.1038/s41598-017-01892-0

[317] Pennisi, E. Why women's bodies abort males during tough times, Dec. 11, 2014 , 3:00 PM https://www.sciencemag.org/news/2014/12/why-women-s-bodies-abort-males-during-tough-times [Accessed 2021.03.10]

[318] Sadeghi, M. R. Unexplained infertility, the controversial matter in management of infertile couples. J Reprod Infertil. 2015;16(1):1–2.

[319] Lie, M. Ravn, M. N., Spilker, K. (2011) Reproductive Imaginations: Stories of Egg and Sperm, NORA – Nordic Journal of Feminist and Gender Research, 19:4, 231-248, DOI: 10.1080/08038740.2011.618463

[320] The First Person Who Ever Saw Sperm Cells Collected Them From His Wife. https://gizmodo.com/the-first-time-anyone-saw-sperm-1708170526 [access 2020-11-12]

[321] How Sperm 'Swim' May Be Nothing But an Optical Illusion. https://gizmodo.com/how-sperm-swim-may-be-nothing-but-an-optical-illusion-1844571256. [access 2020-11-12]

[322] Robertson, S. A., Moldenhauer, L. M. (2014). Immunological determinants of implantation success. International Journal of Developmental Biology, 58(2–4), 205–217. https://doi.org/10.1387/ijdb.140096sr

References

[323] A woman's eggs choose lucky sperm during last moments of conception, study finds https://edition.cnn.com/2020/06/09/health/sperm-choice-female-eggs-wellness/index.html

[324] Li, S., Winuthayanon, W. (2017). Oviduct: Roles in fertilization and early embryo development. Journal of Endocrinology, 232(1), R1–R26. https://doi.org/10.1530/JOE-16-0302

[325] Ghazal, S., Kulp Makarov, J., et al. Egg Transport and Fertilization. Glob. libr. women's med., (ISSN: 1756-2228) 2014; DOI 10.3843/GLOWM.10317

[326] Age and fertility. A guide for patients, 2012. American Society for Reproductive Medicine, Patient Information Series. https://www.reproductivefacts.org/globalassets/rf/news-and-publications/bookletsfact-sheets/english-fact-sheets-and-info-booklets/Age_and_Fertility.pdf

[327] Pfeifer, S., Butts, S., Fossum, G., Gracia, C., La Barbera, A., Mersereau, J., et al. (2017). Optimizing natural fertility: a committee opinion. Fertility and Sterility, 107(1), 52–58. https://doi.org/10.1016/j.fertnstert.2016.09.029

[328] Wilcox, A., Weinberg, C., Baird, D. Timing of sexual intercourse in relation to ovulation. Effects on the Probability of Conception, Survival of the Pregnancy, and Sex of the Baby, The New England Journal of Medicine, 333(23), 1995. DOI:10.1097/0000625.4-199606000-00016

[329] Is a pregnant woman's chance of giving birth to a boy 50 percent? https://www.scientificamerican.com/article/is-a-pregnant-womans-chan/ [access 2020-11-12]

[330] Czeizel, A. E. Periconceptional folic acid containing multivitamin supplementation. European Journal of Obstetrics and Gynecology and Reproductive Biology, Volume 78, Issue 2, 151 – 161

[331] Czeizel, A. E., Dudás, I., Vereczkey, A., Bánhidy, F. (2013). Folate deficiency and folic acid supplementation: The prevention of neural-tube defects and congenital heart defects. Nutrients, 5(11), 4760–4775. https://doi.org/10.3390/nu5114760

[332] Imbard, A., Benoist, J. F., Blom, H. J. Neural tube defects, folic acid and methylation. Int J Environ Res Public Health. 2013;10(9):4352-4389. Published 2013 Sep 17. doi:10.3390/ijerph10094352

[333] Boivin, J., Griffiths, E., Venetis, C. A. Emotional distress in infertile women and failure of assisted reproductive technologies: meta-analysis of prospective psychosocial studies BMJ 2011; 342 :d223

[334] Dahlberg, J., Andersson, G. Fecundity and human birth seasonality in Sweden: a register-based study. Reproductive Health (2019) 16:87 https://doi.org/10.1186/s12978-019-0754-1

[335] Lam, D. A., Miron, J. A. (1994). Global Patterns of Seasonal Variation in Human Fertility. Annals of the New York Academy of Sciences, 709(1), 9–28. https://doi.org/10.1111/j.1749-6632.1994.tb30385.x

[336] Levitas, E., Lunenfeld, E., Weisz, N., et al. Seasonal variations of human sperm cells among 6455 semen samples: a plausible explanation of a seasonal birth pattern. Am J Obstet Gynecol 2013;208:406.e1-6

[337] Potts, M., Campbell, M. History of Contraception. Glob. libr. women's med., (ISSN: 1756-2228) 2009; DOI 10.3843/GLOWM.10376.

[338] Khan, F., Mukhtar, S., Dickinson, I. K., Sriprasad, S. The story of the condom. Indian J Urol. 2013;29(1):12-15.

doi:10.4103/0970-1591.109976

[339] Thiery, M. Gabriele Fallopio (1523–1562) and the Fallopian tube. Gynecol Surg 6, 93–95 (2009). https://doi.org/10.1007/s10397-008-0453-3

[340] Bernstein, A., Jones, K., The Economic Effects of Contraceptive Access: A Review of the Evidence, (IWPR #B381). https://iwpr.org/iwpr-issues/reproductive-health/the-economic-effects-of-contraceptive-access-a-review-of-the-evidence-fact-sheet/ [access 2020-11-12]

[341] Lanzola, E. L., Ketvertis, K. Intrauterine Device. 2020, StatPearls Publishing LLC.

[342] Dean, G., Schwarz, E. B. (2011). "Intrauterine contraceptives (IUCs)". In Hatcher, Robert A.; Trussell, James; Nelson, Anita L.; Cates, Willard Jr.; Kowal, Deborah; Policar, Michael S. (eds.). Contraceptive technology (20th revised ed.). New York: Ardent Media. pp. 147–191.

[343] Sivin, I. IUDs are contraceptives, not abortifacients: a comment on research and belief. Stud Fam Plann. 1989 Nov-Dec;20(6 Pt 1):355-9.

[344] Liao, P. V., Dollin, J. Half a century of the oral contraceptive pill: historical review and view to the future. Can Fam Physician. 2012;58(12):e757-e760.

[317] Bernstein, A., Jones, K. The Economic Effects of Contraceptive Access: A Review of the Evidence. https://iwpr.org/iwpr-issues/reproductive-health/the-economic-effects-of-contraceptive-access-a-review-of-the-evidence-fact-sheet/ [access 2020-11-12]

[346] Cleland, J., Conde-Agudelo, A., Peterson, H., Ross, J., Tsui, A. (2012). Contraception and health. The Lancet, 380(9837), 149–156. https://doi.org/10.1016/S0140-6736(12)60609-6

[347] Fritz, M.A., Speroff, L. Clinical Gynecologic Endocrinology and Infertility. Eighth edition, Lippincott Williams & Wilkins, 2011

[348] Androgenic, corticoid, mineralocorticoid and estrogenic receptors

[349] Grikšiene, R., Ruksenas, O. Effects of hormonal contraceptives on mental rotation and verbal fluency. Psychoneuroendocrinology. Volume 36, Issue 8, September 2011, Pages 1239-1248

[350] Fritz, M.A. and Speroff, L. Clinical Gynecologic Endocrinology and Infertility.Eighth edition, Lippincott Williams & Wilkins, 2011

[351] Ibid.

[352] McDaid, A., Logette, E., Buchillier, V., Muriset, M., Suchon, P., Pache, T. D., et al. (2017) Risk prediction of developing venous thrombosis in combined oral contraceptive users. PLoS ONE 12(7): e0182041. https://doi.org/10.1371/journal.pone.0182041 [access 2020-11-12]

[353] Collaborative Group on Epidemiological Studies on Endometrial Cancer. Endometrial cancer and oral contraceptives: an individual participant meta-analysis of 27 276 women with endometrial cancer from 36 epidemiological studies. Lancet Oncol. 2015 Sep;16(9):1061-1070. doi: 10.1016/S1470-2045(15)00212-0. Epub 2015 Aug 4.

[354] Iversen, L., Fielding, S., Lidegaard, Ø., Mørch, L. S., Skovlund, C.. W., Hannaford, P. C. et al. Association

References

between contemporary hormonal contraception and ovarian cancer in women of reproductive age in Denmark: prospective, nationwide cohort study BMJ 2018; 362 :k3609

[355] Alvergne, A., Högqvist Tabor, V. (2018). Is Female Health Cyclical? Evolutionary Perspectives on Menstruation. Trends in Ecology and Evolution, 33(6), 399–414. https://doi.org/10.1016/j.tree.2018.03.006

[356] Human papillomavirus (HPV) and cervical cancer. https://www.who.int/news-room/fact-sheets/detail/human-papillomavirus-(hpv)-and-cervical-cancer

[357] Bäckström, T., Andreen, L., Birzniece, V., Björn, I., Johansson, I. M., Nordenstam-Haghjo, M., et al.. (2003). The role of hormones and hormonal treatments in premenstrual syndrome. CNS Drugs, 17(5), 325–342. https://doi.org/10.2165/00023210-200317050-00003

[358] Lewis, C. A., Kimmig, A. S., Zsido, R. G., Jank, A., Derntl, B., Sacher, J. Effects of Hormonal Contraceptives on Mood: A Focus on Emotion Recognition and Reactivity, Reward Processing, and Stress Response. Curr Psychiatry Rep. 2019;21(11):115. Published 2019 Nov 7. doi:10.1007/s11920-019-1095-z

[359] Skovlund, C. W., Mørch, L. S., Kessing, L. V., Lidegaard, Ø. Association of Hormonal Contraception With Depression. JAMA Psychiatry. 2016;73:1154–62.

[360] Lundin, C., Danielsson, K. G., Bixo, M., Moby, L., Bengtsdotter, H., Jawad, I., et al. Combined oral contraceptive use is associated with both improvement and worsening of mood in the different phases of the treatment cycle-A double-blind, placebo-controlled randomized trial. Psychoneuroendocrinology. 2017;76:135–43.

[361] Lewis, C. A., Kimmig, A. S., Zsido, R. G., Jank, A., Derntl, B., Sacher, J. Effects of Hormonal Contraceptives on Mood: A Focus on Emotion Recognition and Reactivity, Reward Processing, and Stress Response. Curr Psychiatry Rep. 2019;21(11):115. Published 2019 Nov 7. doi:10.1007/s11920-019-1095-z

[362] Casado-Espada, N. M., de Alarcón, R., de la Iglesia-Larrad, J. I., Bote-Bonaechea, B., Montejo, Á. L. Hormonal Contraceptives, Female Sexual Dysfunction, and Managing Strategies: A Review. J Clin Med. 2019;8(6):908. Published 2019 Jun 25. doi:10.3390/jcm8060908

[363] Scheele, D., Plota, J., Stoffel-Wagner, B., Maier, W., Hurlemann, R. Hormonal contraceptives suppress oxytocin-induced brain reward responses to the partner's face. Soc Cogn Affect Neurosci. 2016;11(5):767–774. doi:10.1093/scan/nsv157

[364] Haselton, M. Hormonal, 2018, Oneworld publications

[365] Casado-Espada, N. M., de Alarcón, R., de la Iglesia-Larrad, J. I., Bote-Bonaechea, B., Montejo, Á. L. Hormonal Contraceptives, Female Sexual Dysfunction, and Managing Strategies: A Review. J Clin Med. 2019;8(6):908. Published 2019 Jun 25. doi:10.3390/jcm8060908

[366] de Castro Coelho, F., Barros, C. (2019). The Potential of Hormonal Contraception to Influence Female Sexuality. International Journal of Reproductive Medicine, 2019, 1–9. https://doi.org/10.1155/2019/9701384

[367] Lewis, C. A., Kimmig, A. S., Zsido, R.G., Jank, A., Derntl, B., Sacher, J. Effects of Hormonal Contraceptives on Mood: A Focus on Emotion Recognition and Reactivity, Reward Processing, and Stress Response. Curr Psychiatry Rep. 2019;21(11):115. Published 2019 Nov 7. doi:10.1007/s11920-019-1095-z

[368] Hamstra, D. A., de Kloet, E. R., van Hemert, A.M., de Rijk, R.H., Van der Does, A. J. Mineralocorticoid receptor haplotype, oral contraceptives and emotional information processing. Neuroscience. 2015;286:412–22. https://doi.org/10.1016/j.neuroscience.2014.12.004.

[369] CGEI

[370] Williams Obstetrics, 2th edition. McGraw-Hill Education, New York, USA

[371] Zhi-Kun Li, Le-Yun Wang, Li-Bin Wang, Gui-Hai Feng, Xue-Wei Yuan, Chao Liu, Kai Xu, Yu-Huan Li, Hai-Feng Wan. Ying Zhang, Yu-Fei Li, Xin Li, Wei Li, Qi Zhou, Bao-Yang Hu. Generation of Bimaternal and Bipaternal Mice from Hypomethylated Haploid ESCs with Imprinting Region Deletions. Cell Stem Cell 23 (5), P665-676.E4, Nov 1, 2018

[372] Lorenz, T. K., Worthman, C. M., Vitzthum, V. J. (2015). Links among inflammation, sexual activity and ovulation: Evolutionary trade-offs and clinical implications. Evolution, Medicine and Public Health, 2015(1). https://doi.org/10.1093/emph/eov029

[373] How women can deal with periods in space. https://phys.org/news/2016-04-women-periods-space.html [Access 2021.03.11]

[374] Fritz, M.A., Speroff, L. Clinical Gynecologic Endocrinology and Infertility. Eighth edition, Lippincott Williams & Wilkins, 2011

[375] Liu, Y., Gold, E. B., Lasley, B. L. Johnson, W. O. Factors Affecting Menstrual Cycle Characteristics, American Journal of Epidemiology, Volume 160, Issue 2, 15 July 2004, Pages 131–140, https://doi.org/10.1093/aje/kwh188

[376] Lord, A. M. "The Great Arcana of the Deity": Menstruation and Menstrual Disorders in Eighteenth-Century British Medical Thought." Bulletin of the History of Medicine, vol. 73 no. 1, 1999, p. 38-63. Project MUSE, doi:10.1353/bhm.1999.0036.

[377] McNeil, J., Doucet, É. (2012). Possible factors for altered energy balance across the menstrual cycle: A closer look at the severity of PMS, reward driven behaviors and leptin variations. European Journal of Obstetrics and Gynecology and Reproductive Biology, 163(1), 5–10. https://doi.org/10.1016/j.ejogrb.2012.03.008

[378] Fritz, M.A., Speroff, L. Clinical Gynecologic Endocrinology and Infertility. Eighth edition, Lippincott Williams & Wilkins, 2011

[379] Meczekalski, B., Katulski, K., Czyzyk, A., Podfigurna-Stopa, A., Maclejewska-Jeske, M. (2014). Functional hypothalamic amenorrhea and its influence on women's health. Academic Psychiatry, 37(11), 1049–1056. https://doi.org/10.1007/s40618-014-0169-3

[380] Shufelt, C. L., Torbati, T., Dutra, E. B. (2017). Hypothalamic Amenorrhea and the Long-Term Health Consequences. Semin Reprod Med., 35(3), 256–262. https://doi.org/10.1016/j.physbeh.2017.03.040

[381] Kalantaridou, S. N., Makrigiannakis, A., Zoumakis, E., Chrousos, G. P. (2004). Stress and the female reproductive system. Journal of Reproductive Immunology, 62(1), 61–68.

[382] Ibid.

References

[383] Fritz, M.A., Speroff, L. Clinical Gynecologic Endocrinology and Infertility. Eighth edition, Lippincott Williams & Wilkins, 2011

[384] Shufelt, C. L., Torbati, T. B. S., Dutra, E. B. (2017). Hypothalamic Amenorrhea and the Long-Term Health Consequences. Semin Reprod Med., 35(3), 256–262. https://doi.org/10.1016/j.physbeh.2017.03.040

[385] Richardson, V. R., Smith, K. A., Carter, A. M., Adipose tissue inflammation: feeding the development of type 2 diabetes mellitus. Immunobiology. 2013 Dec;218(12):1497-504. doi: 10.1016/j.imbio.2013.05.002. Epub 2013 May 16.

[386] Dunaif, A., Marshall, J. (2013). All Women with PCOS Should Be Treated For Insulin Resistance. National Institute of Health, 97(1), 18–22. https://doi.org/10.1016/j.fertnstert.2011.11.036.All

[387] Committee, P., & Society, A. (2015). Obesity and reproduction: A committee opinion. Fertility and Sterility, 104(5), 1116–1126. https://doi.org/10.1016/j.fertnstert.2015.08.018

[388] Silvestris, E., de Pergola, G., Rosania, R., Loverro, G. (2018). Obesity as disruptor of the female fertility. Reproductive Biology and Endocrinology, 16(1), 1–13. https://doi.org/10.1186/s12958-018-0336-z

[389] Committee, P., & Society, A. (2015). Obesity and reproduction: A committee opinion. Fertility and Sterility, 104(5), 1116–1126. https://doi.org/10.1016/j.fertnstert.2015.08.018

[390] Chang, R., Kazer, R. Polycystic Ovary Syndrome, Glob. libr. women's med., (ISSN: 1756-2228) 2014; DOI 10.3843/GLOWM.10301

[391] Fauser, B. C. J. M., Tarlatzis, B. C., Rebar, R. W., Legro, R. S., Balen, A. H., Lobo, R., et al. (2012). Consensus on women's health aspects of polycystic ovary syndrome (PCOS). Human Reproduction, 27(1), 14–24. https://doi.org/10.1093/humrep/der396

[392] Garg, D., Tal, R. The role of AMH in the pathophysiology of polycystic ovarian syndrome. Review| Vol 33, Issue 1, P15-28, , 2016, DOI:https://doi.org/10.1016/j.rbmo.2016.04.007

[393] Jamil, Z., Fatima, S. S., Ahmed, K., Malik, R. (2016). Anti-Müllerian Hormone: Above and beyond Conventional Ovarian Reserve Markers. Disease Markers, 2016. https://doi.org/10.1155/2016/5246217

[394] Ndefo, U. A., Eaton, A., Green, M. R. (2013). Polycystic ovary syndrome: A review of treatment options with a focus on pharmacological approaches. P and T, 38(6), 336–355.

[395] Tata, B., Mimouni, N. E. H., Barbotin, A-L., et al (2018) Elevated prenatal anti-Müllerian hormone reprograms the fetus and induces polycystic ovary syndrome in adulthood. Nature Medicine 24:834–846. https://doi.org/10.1038/s41591-018-0035-5

[396] Nestler, J. E., Jakubowicz, D. J. (1997). Lean women with polycystic ovary syndrome respond to insulin reduction with decreases in ovarian P450c17α activity and serum androgens. Journal of Clinical Endocrinology and Metabolism, 82(12), 4075–4079. https://doi.org/10.1210/jcem.82.12.4431

[397] Dunaif, A., Marshall, J. (2013). All Women with PCOS Should Be Treated For Insulin Resistance. National Institute of Health, 97(1), 18–22. https://doi.org/10.1016/j.fertnstert.2011.11.036.All

[398] Hirschberg, A. L. Female hyperandrogenism and elite sport [published online ahead of print, 2020 Mar 1]. Endocr Connect. 2020;9(4):R81-R92. doi:10.1530/EC-19-0537

[399] Lebbi, I., Ben Temime, R., Fadhlaoui, A., Feki, A. Ovarian Drilling in PCOS: Is it Really Useful?. Front Surg. 2015;2:30. Published 2015 Jul 17. doi:10.3389/fsurg.2015.00030

[400] Bastos Maia, S., Rolland Souza, A. S., Costa Caminha, M. F., et al. Vitamin A and Pregnancy: A Narrative Review. Nutrients. 2019;11(3):681. Published 2019 Mar 22. doi:10.3390/nu11030681

[401] Hager, M., Wenzl, R., Riesenhuber, S., et al. The Prevalence of Incidental Endometriosis in Women Undergoing Laparoscopic Ovarian Drilling for Clomiphene-Resistant Polycystic Ovary Syndrome: A Retrospective Cohort Study and Meta-Analysis. J Clin Med. 2019;8(8):1210. Published 2019 Aug 14. doi:10.3390/jcm8081210

[402] Taraborrelli, S. (2015). Physiology, production and action of progesterone. Acta Obstetricia et Gynecologica Scandinavica, 94, 8–16. https://doi.org/10.1111/aogs.12771

[403] Miller, P, Soules, M, Glob. libr. women's med., (ISSN: 1756-2228) 2009; DOI 10.3843/GLOWM.10327

[404] Simpson, E. R., Rochelle, D.B., Carr, B. R., Mac Donald, P. C. (1980) Plasma lipoproteins in follicular fluid of human ovaries. J Clin Endocrinol Metab 51: 1469-1471.

[405] Mesen, T. B., Young, S. L. (2015). Progesteron and the Luteal Phase. Obstetrics and Gynecology Clinics of North America, 43(1), 135–151. doi: 10.1016/j.ogc.2014.10.003.

[406] American Thyroid Association) Prevalence and Impact of Thyroid Disease. https://www.thyroid.org/media-main/press-room/

[407] Thyroid disease. https://www.womenshealth.gov/a-z-topics/thyroid-disease [access 2020-11-12]

[408] Guyton, A. C., Hall, J. E. (2006). Medical physiology 11th edition. Textbook of Medical Physiology.

[409] Bachelot, A., Binart, N. Reproductive role of prolactin. Reproduction. 2007 Feb 1;133(2):361-9.

[410] Marques, P., Skorupskaite, K., George, J. T., et al. Physiology of GNRH and Gonadotropin Secretion. [Updated 2018 Jun 19]. In: Feingold KR, Anawalt B, Boyce A, et al., editors. Endotext [Internet]. South Dartmouth (MA): MDText.com, Inc.; 2000-. Available from: https://www.ncbi.nlm.nih.gov/books/NBK279070/

[411] https://www.nytimes.com/2012/07/18/science/fda-bans-bpa-from-baby-bottles-and-sippy-cups.html

[412] Mirmira, P; Evans-Molina, C (2014). "Bisphenol A, obesity, and type 2 diabetes mellitus: genuine concern or unnecessary preoccupation?". Translational Research (Review). 164 (1): 13–21. doi:10.1016/j.trsl.2014.03.003. hdl:1805/8373. PMC 4058392. PMID 24686036.

[413] Gore, A. C., Chappell, V. A., Fenton, S. E., et al. Executive Summary to EDC-2: The Endocrine Society's Second Scientific Statement on Endocrine-Disrupting Chemicals. Endocr Rev. 2015;36(6):593–602. doi:10.1210/er.2015-1093

[414] Okada, H., Tokunaga, T., Liu, X., Takayanagi, S., Matsushima, A., Shimohigashi, Y. (2008). Direct evidence revealing structural elements essential for the high binding ability of bisphenol a to human estrogen-related receptor-γ. Environmental Health Perspectives, 116(1), 32–38. https://doi.org/10.1289/ehp.10587

References

[415] Gore, A. C., Chappell, V. A., Fenton, S. E., et al. Executive Summary to EDC-2: The Endocrine Society's Second Scientific Statement on Endocrine-Disrupting Chemicals. Endocr Rev. 2015;36(6):593–602. doi:10.1210/er.2015-1093

[416] La Merrill, M.A., Vandenberg, L.N., Smith, M.T. et al. Consensus on the key characteristics of endocrine-disrupting chemicals as a basis for hazard identification. Nat Rev Endocrinol 16, 45–57 (2020). https://doi.org/10.1038/s41574-019-0273-8

[417] Eberle, C. E., Sandler, D. P., Taylor, K. W., White, A. J. Hair dye and chemical straightener use and breast cancer risk in a large US population of black and white women. , .Int J Cancer. 2020 Jul 15;147(2):383-391 https://doi.org/10.1002/ijc.32738

[418] Breast Cancer Risk Factors. https://www.breastcancer.org/symptoms/understand_bc/risk/factors [access 2020-11-12].

[419] Wee, Sze Yee, Aris, Ahmad Zaharin. (2017). Endocrine disrupting compounds in drinking water supply system and human health risk implication. Environment International. 106. 10.1016/j.envint.2017.05.004.

[420] Iwanowicz, L. R., Blazer, V. S., Pinkney, A. E., Guy, C. P., Major, A. M., Munney. K., et al. Evidence of estrogenic endocrine disruption in smallmouth and largemouth bass inhabiting Northeast U.S. national wildlife refuge waters: A reconnaissance study. Ecotoxicol Environ Saf. 2016 Feb;124:50-59. doi: 10.1016/j.ecoenv.2015.09.035. Epub 2015 Oct 19.

[421] Adeel, M., Song, X., Wang, Y., Francis, D., Yang, Y. Environmental impact of estrogens on human, animal and plant life: A critical review, Environment International Volume 99, February 2017, 107-119.

[422] Wise, A., O'Brien, K., Woodruff, T. Are oral contraceptives a significant contributor to the estrogenicity of drinking water? Environ Sci Technol. 2011 Jan 1;45(1):51-60. doi: 10.1021/es1014482. Epub 2010 Oct 26.

[423] Birth-control pills could add 10 million doses of hormones to our wastewater every day. Some of that estrogen may wind up in our taps. https://www.businessinsider.com/birth-control-pills-hormones-estrogen-drinking-water-health-effects-2019-10?r=US&IR=T [access 2020-12-01]

[424] Endocrine Disrupting Chemicals Are Everywhere, but How Do They Affect Pregnancy? https://www.healthline.com/health-news/what-do-we-know-about-endocrine-disruptors-during-pregnancy [access 2020-12-01]

[425] Philips, E. M., Kahn, L. G., Jaddoe, V. W. V., et al. First Trimester Urinary Bisphenol and Phthalate Concentrations and Time to Pregnancy: A Population-Based Cohort Analysis. J Clin Endocrinol Metab. 2018;103(9):3540–3547. doi:10.1210/jc.2018-00855

[426] Mínguez-Alarcón, L., Gaskins, A. J., Chiu, Y-H., Souter, I., Williams, P. L., Calafat, A. M., Russ Hausera, J. E. C. (2016). Dietary folate intake and modification of the association of urinary bisphenol A concentrations with in vitro fertilization outcomes among women from a fertility clinic. Reproductive Toxicology, 65, 104–112. https://doi.org/10.1016/j.physbeh.2017.03.040

[427] Chavarro, J. E., Mínguez-Alarcón, L., Chiu, Y. H. et al. Soy Intake Modifies the Relation Between Urinary Bisphenol A Concentrations and Pregnancy Outcomes Among Women Undergoing Assisted Reproduction. J

Clin Endocrinol Metab. 2016;101(3):1082–1090. doi:10.1210/jc.2015-3473

[428] Are Phytoestrogens Good for You? https://www.healthline.com/health/phytoestrogens [access 2020-12-01]

[429] Chen, M., Rao, Y., Zheng, Y., Wei, S., Li, Y., Guo, T., Yin, P. (2014). Association between soy isoflavone intake and breast cancer risk for pre- and post-menopausal women: A meta-analysis of epidemiological studies. PLoS ONE, 9(2). https://doi.org/10.1371/journal.pone.0089288

[430] Mitchell, J. H., Cawood, E., Kinniburgh, D., Provan, A., Collins, A. R., Irvine, D. S. Effect of a phytoestrogen food supplement on reproductive health in normal males. Clin Sci (Lond). 2001 Jun;100(6):613-8.

[431] Cederroth, C. R., Auger, J., Zimmermann, C., Eustache, F., Nef, S. Soy, phyto-oestrogens and male reproductive function: a review. Int J Androl. 2010 Apr;33(2):304-16. doi: 10.1111/j.1365-2605.2009.01011.x. Epub 2009 Nov 16.

[432] Mínguez-Alarcón, L., Afeiche, M. C., Chiu, Y. H. et al. Male soy food intake was not associated with in vitro fertilization outcomes among couples attending a fertility center. Andrology. 2015;3(4):702–708. doi:10.1111/andr.12046

[433] Hamilton-Reeves, J. M., Vazquez, G., Duval, S. J., Phipps, W. R., Kurzer, M. S., Messina, M. J. Clinical studies show no effects of soy protein or isoflavones on reproductive hormones in men: results of a meta-analysis. Fertility and Sterility, Volume 94, Issue 3, 997 – 1007

[434] Messina, M. Soybean isoflavone exposure does not have feminizing effects on men: a critical examination of the clinical evidence.Fertil Steril. 2010 May 1;93(7):2095-104. doi: 10.1016/j.fertnstert.2010.03.002. Epub 2010 Apr 8.

[435] Adgent, M. A., Umbach, D. M., Zemel, B. S., Kelly, A., Schall, J. I., Ford, E. G. et al. A longitudinal study of estrogen-responsive tissues and hormone concentrations in infants fed soy formula. The Journal of Clinical Endocrinology & Metabolism, 2018; DOI: 10.1210/jc.2017-02249

[436] Fraser, I. S., Critchley, H. O. D., Munro, M. G., Broder, M. (2007). Can we achieve international agreement on terminologies and definitions used to describe abnormalities of menstrual bleeding? Human Reproduction, 22(3), 635–643. https://doi.org/10.1093/humrep/del478

[437] Bull, J.R., Rowland, S.P., Scherwitzl, E.B. et al. Real-world menstrual cycle characteristics of more than 600,000 menstrual cycles. npj Digit. Med. 2, 83 (2019) doi:10.1038/s41746-019-0152-7

[438] Jeyaseelan, L., Antonisamy, B., Rao, P.S. Pattern of menstrual cycle length in south Indian women: a prospective study. Soc Biol. 1992 Fall-Winter;39(3-4):306-9.

[439] Liu, Y., Gold, E. B., Lasley, B. L., Johnson, W. O. (2004). Factors affecting menstrual cycle characteristics. American Journal of Epidemiology, 160(2), 131–140. https://doi.org/10.1093/aje/kwh188

[440] Santoro, N., Lasley, B., McConnell, D., Allsworth, J., Crawford, S. et al. Body Size and Ethnicity Are Associated with Menstrual Cycle Alterations in Women in the Early Menopausal Transition: The Study of Women's Health across the Nation (SWAN) Daily Hormone Study, The Journal of Clinical Endocrinology & Metabolism, Volume 89, Issue 6, 1 June 2004, Pages 2622–2631, https://doi.org/10.1210/jc.2003-031578

References

[441] Bull, J.R., Rowland, S.P., Scherwitzl, E.B. et al. Real-world menstrual cycle characteristics of more than 600,000 menstrual cycles. npj Digit. Med. 2, 83 (2019) doi:10.1038/s41746-019-0152-7

[442] Carmina, E., Lobo, R. A. Evaluation of Hormonal Status. in Yen & Jaffe's Reproductive Endocrinology (Sixth Edition), 2009

[443] Goodale, B. M., Shilaih, M., Falco, L., Dammeier, F., Hamvas, G., Leeners, B. (2019). Wearable sensors reveal menses-driven changes in physiology and enable prediction of the fertile window: an observational study. Journal of Medical Internet Research, 21(4), e13404.

[444] Stachenfeld, N. S., Silva, C., Keefe, D. L. (2000). Estrogen modifies the temperature effects of progesterone. Journal of Applied Physiology, 88(5), 1643–1649. https://doi.org/10.1152/jappl.2000.88.5.1643

[445] McClintock, M. K. (1971). "Menstrual Synchrony and Suppression". Nature. 229 (5282): 244–5. Bibcode:1971Natur.229..244M. doi:10.1038/229244a0. PMID 4994256.

[446] Harris, A. L.; Vitzthum, V. J. (2013). "Darwin's Legacy: An Evolutionary View of Women's Reproductive and Sexual Functioning". Journal of Sex Research. 50 (3–4): 207–46. doi:10.1080/00224499.2012.763085. PMID 23480070.

[447] Strassmann, B. I. (February 1997). "The Biology of Menstruation in Homo Sapiens: Total Lifetime Menses, Fecundity, and Nonsynchrony in a Natural-Fertility Population". Current Anthropology. 38 (1): 123–129. doi:10.1086/204592. JSTOR 2744446.

[448] Wilcox, A et al. Time of Implantation of the Conceptus and Loss of Pregnancy. June 10, 1999

N Engl J Med 1999; 340:1796-1799D OI: 10.1056/NEJM199906103402304

[449] Fritz, M. A., Speroff, L. Clinical Gynecologic Endocrinology and Infertility, Philadelphia PA: Lippincott Williams & Wilkins, Eighth edition 2011.

[450] Jones, C. J. P., Choudhury, R. H., Aplin. J. D. Tracking nutrient transfer at the human maternofetal interface from 4 weeks to term. Placenta, Volume 36, Issue 4, April 2015, 372-380.

[451] Okada, H., Tsuzuki, T., Murata, H. Decidualization of the human endometrium. Reprod Med Biol. 2018;17(3):220–227. Published 2018 Feb 1. doi:10.1002/rmb2.12088

[452] Cole, L. A. Biological functions of hCG and hCG-related molecules. Reprod Biol Endocrinol. 2010;8:102. doi:10.1186/1477-7827-8-102

[453] Lee, N. M., Saha, S. Nausea and vomiting of pregnancy. Gastroenterol Clin North Am. 2011;40(2):309–vii. doi:10.1016/j.gtc.2011.03.009

[454] Sherman, P. W., Flaxman, S. M. Nausea and vomiting of pregnancy in an evolutionary perspective. Am J Obstet Gynecol. 2002 May;186. doi: 10.1067/mob.2002.122593

[455] Fetal Circulation. https://www.heart.org/en/health-topics/congenital-heart-defects/symptoms--diagnosis-of-congenital-heart-defects/fetal-circulation#.WHGPgPkrLIU [access 2020-12-06]

[456] Kumar, P., Magon, N. Hormones in pregnancy. Niger Med J. 2012;53(4):179–183. doi:10.4103/0300-1652.107549

[457] Guyton, A. C., Hall, J. E. Textbook of Medical Physiology. Philadelphia, PA: Elsevier Saunders, 11th edition 2006....

[458] The Mysterious Tree of a Newborn's Life. https://www.nytimes.com/2014/07/15/health/the-push-to-understand-the-placenta.html [access 2020-12-06]

[459] Pregnancy week by week. https://www.mayoclinic.org/healthy-lifestyle/pregnancy-week-by-week/in-depth/placenta/art-20044425

[460] Labor and delivery, postpartum care. https://www.mayoclinic.org/healthy-lifestyle/labor-and-delivery/expert-answers/eating-the-placenta/faq-20380880 [access 2020-12-06]

[461] Kodogo, V., Azibani, F., Sliwa, K. Role of pregnancy hormones and hormonal interaction on the maternal cardiovascular system: a literature review. Clinical Research in Cardiology (2019) 108:831–846, https://doi.org/10.1007/s00392-019-01441-x

[462] Guyton, A. C., Hall, J. E. Textbook of Medical Physiology. Philadelphia, PA: Elsevier Saunders, 11th edition 2006.

[463] 'My oestrogen levels were all over the place': when men have 'sympathy pregnancies'. https://www.theguardian.com/lifeandstyle/2019/jul/22/my-oestrogen-levels-were-all-over-the-place-when-men-have-sympathy-pregnancies [access 2020-12-06]

[464] What's behind the metallic taste? https://www.healthline.com/health/pregnancy/metallic-taste-in-mouth#causes

[465] Barclay, M, Glob. libr. women's med., (ISSN: 1756-2228) 2009; DOI 10.3843/GLOWM.10103

[466] Human Placental Lactogen: What It Can Tell You About Your Pregnancy https://www.healthline.com/health/human-placental-lactogen [access 2020-12-06]

[467] Dean, L., McEntyre, J. The Genetic Landscape of Diabetes [Internet]. Bethesda (MD): National Center for Biotechnology Information (US); 2004. Chapter 5, Gestational Diabetes. 2004 Jul 7. Available from: https://www.ncbi.nlm.nih.gov/books/NBK1668/

[468] Gestational Diabetes and Pregnancy. https://www.cdc.gov/pregnancy/diabetes-gestational.html [access 2020-12-06].

[469] Guyton, A. C., Hall, J. E. Textbook of Medical Physiology. Philadelphia, PA: Elsevier Saunders, 11th edition 2006.

[470] Hormones of pregnancy and labour https://www.yourhormones.info/topical-issues/hormones-of-pregnancy-and-labour/ [acces 2020-12-06

[471] Duthie, L., Duthie, L., Reynolds, R. M.: Changes in the Maternal Hypothalamic-Pituitary-Adrenal Axis in Pregnancy and Postpartum: Influences on Maternal and Fetal Outcomes. Neuroendocrinology 2013;98:106-115. doi: 10.1159/000354702

[472] Stress hormones during pregnancy https://www.parentingscience.com/Stress-hormones-during-pregnancy.

References

html [access 2020-12-06]

[473] McLean and Smith 1999. Corticotropin-releasing Hormone in Human Pregnancy and Parturition Trends. Endocrinol Metab 10(5):174-178.

[474] Majzoub, J. A., Karalis, K. P. 1999. Placental corticotrophin-releasing hormone: Function and regulation. Am J Obstet Gynecol. 180:S242-246.

[475] McLean and Smith 2001. Corticotropin-releasing Hormone and human parturition. Reproduction 121: 493-501.

[476] Crowley, P. 2000. Prophylactic corticosteroids for preterm birth. Cochrane Database of Systematic Review, issue 2, CD000065.

[477] Matthews, S. G., Owen, D., Kalabis, G., Banjamin, S., Setiawan, E. B., Dunn, E. A., Andrews, M. H. 2004. Fetal glucocorticoid exposure and hypothalamo-pituitary-adrenal (HPA) function after birth. Endocrine Research 30(4): 827-836.

[478] Bardi, M., French, J. A., Ramirez, M., Brent, B. 2004. The role of the endocrine system in baboon maternal behavior. Biol Psychiatry 55: 724-732.

[479] Stallings, J., Fleming, A. S., Corter, C., Worthman, C., Steiner, M. 2001. The effects of infant cries and odors on sympathy, cortisol, and autonomic responses in new mothers and nonpostpartum women. Parenting: Science and Practice 1: 71-100.

[480] Mastripieri, D. 1999. The biology of human parenting: Insights from nonhuman primates. Neuroscience and biohavioral reviews 23:411-422.

[481] Power, M. L., Shulkin, J. 2006. Functions of corticotrophin-releasing hormone in anthropoid primates: from brain to placenta. Am J Hum Biol. 18(4): 431-447.

[482] Guyton, A. C., Hall, J. E. Textbook of Medical Physiology. Philadelphia, PA: Elsevier Saunders, 11th edition 2006.

[483] Kodogo, V., Azibani, F., Sliwa, K. Role of pregnancy hormones and hormonal interaction on the maternal cardiovascular system: a literature review. Clinical Research in Cardiology (2019) 108:831–846, https://doi.org/10.1007/s00392-019-01441-x

[484] Barclay, M, Glob. libr. women's med., (ISSN: 1756-2228) 2009; DOI 10.3843/GLOWM.10103

[485] Loerup, L., Pullon, R.M., Birks, J. et al. Trends of blood pressure and heart rate in normal pregnancies: a systematic review and meta-analysis. BMC Med 17, 167 (2019). https://doi.org/10.1186/s12916-019-1399-1

[486] Sanghavi, M., Rutherford, J. D., Cardiovascular Physiology of Pregnancy, Circulation. 2014;130:1003–1008 https://doi.org/10.1161/CIRCULATIONAHA.114.009029

[487] Braunthal, S., Brateanu, A. Hypertension in pregnancy: Pathophysiology and treatment. SAGE Open Med. 2019;7:2050312119843700. Published 2019 Apr 10. doi:10.1177/2050312119843700

[488] Al-Jameil, N., Aziz Khan, F., Fareed Khan, M., Tabassum, H. A brief overview of preeclampsia. J Clin Med

Res. 2014;6(1):1–7. doi:10.4021/jocmr1682w

[489] Eclampsia https://www.healthline.com/health/eclampsia [access 2020-12-06]

[490] Podcast: Zombified. Placental hijacking: Harvard evolutionary biologist David Haig being interviewed by Athena Aktipis

[491] Galaviz-Hernandez, C. et al. Paternal Determinants in Preeclampsia. Front. Physiol., 07 January 2019 | https://doi.org/10.3389/fphys.2018.01870

[492] Nagayama, S., Ohkuchi, A., Usui, R., Matsubara, S., Suzuki, M. (2014) The Role of the Father in the Occurrence of Preeclampsia. Med J ObstetGynecol 2(2): 1029.

[493] Roser, M., Ritchie, H. (2013) – "Maternal Mortality". Published online at OurWorldInData.org. Retrieved from: 'https://ourworldindata.org/maternal-mortality' [Online Resource]

[494] Pregnancy Mortality Surveillance System. https://www.cdc.gov/reproductivehealth/maternal-mortality/pregnancy-mortality-surveillance-system.htm [acc ess 2020.11.22]

[495] Hoyert, D. L., Miniño, A M. Maternal Mortality in the United States: Changes in Coding, Publication, and Data Release, 2018, National Vital Statistics Reports, Volume 69, Number 2 January 30, 2020

[496] Birth and care of young. https://seaworld.org/animals/all-about/polar-bear/care-of-young/ [access 2020-12-06]

[497] Weight gain during pregnancy. Committee Opinion No. 548. American College of Obstetricians and Gynecologists. Obstet Gynecol 2013;121:210–2.

[498] Thompson, J., Irgens, L., Skjaerven, R., Rasmussen, S. Placenta weight percentile curves for singleton deliveries. BJOG 2007;114:715–720.

[499] Fischer, R, Glob. libr. women's med., (ISSN: 1756-2228) 2008; DOI 10.3843/GLOWM.10208

[500] Guyton, A. C. Hall, J. E. Textbook of Medical Physiology. Philadelphia, PA: Elsevier Saunders, 11th edition 2006.

[501] Ibid.

[502] Guyton, A. C., Hall, J. E. Textbook of Medical Physiology. Philadelphia, PA: Elsevier Saunders, 11th edition 2006.

[503] Almaghamsi, A., Almalki, M. H., Buhary, B. M. Hypocalcemia in Pregnancy: A Clinical Review Update. Oman Med J. 2018;33(6):453-462. doi:10.5001/omj.2018.85

[504] Pregnancy week by week https://www.mayoclinic.org/healthy-lifestyle/pregnancy-week-by-week/in-depth/anemia-during-pregnancy/art-20114455 [accessed 2020.11.22]

[505] Institute of Medicine (US) and National Research Council (US) Committee to Reexamine IOM Pregnancy Weight Guidelines; Rasmussen KM, Yaktine AL, editors. Weight Gain During Pregnancy: Reexamining the Guidelines. Washington (DC): National Academies Press (US); 2009. 5, Consequences of Gestational Weight Gain for the Mother. Available from: https://www.ncbi.nlm.nih.gov/books/NBK32818/

References

506 Barclay, M, Glob. libr. women's med., (ISSN: 1756-2228) 2009; DOI 10.3843/GLOWM.10103

507 The pelvic floor is full of surprises. Here's what you need to know https://www.healthline.com/health/parenting/your-pelvic-floor-explained#The-pelvic-floor-is-full-of-surprises.-Heres-what-you-need-to-know   [access 2020-12-06]

508 Giarenis, I., Robinson, D. Prevention and management of pelvic organ prolapse. F1000Prime Rep. 2014;6:77. Published 2014 Sep 4. doi:10.12703/P6-77

509 Fonti, Y., Giordano, R., Cacciatore, A., Romano, M., La Rosa, B. Post partum pelvic floor changes. J Prenat Med. 2009;3(4):57–59.

510 Fonti, Y., Giordano, R., Cacciatore, A., Romano, M., La Rosa, B. Post partum pelvic floor changes. J Prenat Med. 2009;3(4):57–59.

511 Fischer, R, Glob. libr. women's med., (ISSN: 1756-2228) 2008; DOI 10.3843/GLOWM.10208

512 Amniocentesis (amniotic fluid test). https://medlineplus.gov/lab-tests/amniocentesis-amniotic-fluid-test/ [access 2020-12-06]

513 https://parenting.nytimes.com/health/fatherhood-mens-bodies?smid=tw-nytimes&smtyp=cur [access 2020-12-06]

514 Pregnancy week by week. https://www.mayoclinic.org/healthy-lifestyle/pregnancy-week-by-week/expert-answers/baby-brain/faq-20057896 [access 2020-12-06]

515 Chan, W. F., Nelson, J. L. Microchimerism in the human brain: more questions than answers. Chimerism. 2013;4(1):32–33. doi:10.4161/chim.24072

516 Rijnink, E. C., Penning, M. E., Wolterbeek, R., Wilhelmus, S., Zandbergen, M., van Duinen S. G., et al. Tissue microchimerism is increased during pregnancy: a human autopsy study, Molecular Human Reproduction, Volume 21, Issue 11, November 2015, Pages 857–864, https://doi.org/10.1093/molehr/gav047

517 Labor, S., Maguire, S. The pain of labor. Reviews in Pain 2 (2), 2008

518 Guyton, A. C., Hall JE. Textbook of Medical Physiology Eleventh edition. Elsevier Inc 2006 Philadelphia, Pennsylvania

519 Williams Obstetrics, 24th edition. McGraw-Hill Education, New York, USA

520 Ibid.

521 Natural Ways to Induce Labor. https://www.healthline.com/health/pregnancy/natural-ways-to-induce-labor [access 2020-12-06]

522 Fox, N., Gelber, S., Chasen, S. Physical And Sexual Activity During Pregnancy Are Not Associated With The Onset Of Labor And Mode Of Delivery In Low Risk Term Nulliparous Women. The Internet Journal of Gynecology and Obstetrics. 2006. Volume 8 Number 1.

523 Oxytocin. https://www.yourhormones.info/hormones/oxytocin/ [access 2020-12-06]

524 Ravanos, K., Dagklis, T., Petousis, S., Margioula-Siarkou, C., Prapas, Y., Prapas, N. (2015) Factors implicated

in the initiation of human parturition in term and preterm labor: a review, Gynecological Endocrinology, 31:9, 679-683.

[525] Gao, L., Rabbitt, E. H., Condon, J. C., Renthal, N. E., Johnston, J. M., et al. Steroid receptor coactivators 1 and 2 mediate fetal-to-maternal signaling that initiates parturition. Journal of Clinical Investigation, 2015; DOI: 10.1172/JCI78544

[526] Placenta 'Switch' that Kickstarts Labor May Solve Long-Standing Mystery https://www.livescience.com/51983-placenta-switch-starts-labor.html [access 2020-12-06]

[527] Ravanos, K., Dagklis, T., Petousis, S., Margioula-Siarkou, C., Prapas, Y., Prapas, N. (2015) Factors implicated in the initiation of human parturition in term and preterm labor: a review, Gynecological Endocrinology, 31:9, 679-683.

[528] Guyton, A. C., Hall, J. E. Textbook of Medical Physiology. Philadelphia, PA: Elsevier Saunders, 11th edition 2006.

[529] Ravanos, K., Dagklis, T., Petousis, S., Margioula-Siarkou, C., Prapas, Y., Prapas, N. (2015) Factors implicated in the initiation of human parturition in term and preterm labor: a review, Gynecological Endocrinology, 31:9, 679-683.

[530] Placenta 'Switch' that Kickstarts Labor May Solve Long-Standing Mystery https://www.livescience.com/51983-placenta-switch-starts-labor.html [access 2020-12-06]

[531] During Coronavirus Lockdowns, Some Doctors Wondered: Where Are the Preemies? https://www.nytimes.com/2020/07/19/health/coronavirus-premature-birth.html [access 2020-12-06]

[532] Cervix Dilation Chart: The Stages of Labor. https://www.healthline.com/health/pregnancy/cervix-dilation-chart [access 2020-12-06]

[533] Williams Obstetrics, 24th edition. McGraw-Hill Education, New York, USA...

[534] Labor, S., Maguire, S. The Pain of Labour. Rev Pain. 2008;2(2):15-19. doi:10.1177/204946370800200205

[535] Ibid.

[536] Pain relief in labour. https://www.nhs.uk/conditions/pregnancy-and-baby/pain-relief-labour/ [access 2020-12-06]

[537] Wells, H.. https://www.britannica.com/biography/Horace-Wells [access 2020-12-06]

[538] Guyton, A. C., Hall, J. E. Textbook of Medical Physiology Eleventh edition. Elsevier Inc 2006 Philadelphia, Pennsylvania

[539] Williams Obstetrics, 24th edition. McGraw-Hill Education, New York, USA

[540] Guyton, A. C., Hall, J. E. Textbook of Medical Physiology Eleventh edition. Elsevier Inc 2006 Philadelphia, Pennsylvania

[541] Williams Obstetrics, 24th edition. McGraw-Hill Education, New York, USA

[542] Vain, N. E., Satragno, D. S., Gorenstein, A. N., Gordillo, J. E., Berazategui, J. P., Alda, G., Prudent, L. M. Effect

References

of gravity on volume of placental transfusion: a multicentre, random, non-inferiority trial. The Lancet, 384, 9939, 235-240, 2014. https://doi.org/10.1016/S0140-6736(14)60197-5

[543] Placenta accrete. https://www.mayoclinic.org/diseases-conditions/placenta-accreta/symptoms-causes/syc-20376431 [access 2020-12-06]

[544] Jangsten, E., Mattsson, L., Lyckestam, I., Hellström, A., Berg, M. A comparison of active management and expectant management of the third stage of labour: a Swedish randomised controlled trial. BJOG 2011;118:362–369.

[545] Chimpanzee "maternity leave" might save newborn babies from cannibalism https://www.earthtouchnews.com/natural-world/animal-behaviour/chimpanzee-maternity-leave-might-save-newborn-babies-from-cannibalism/ [access 2020-12-06]

[546] Hoekzema, E., Barba-Müller (E., Pozzobon, C. et al. Pregnancy leads to long-lasting changes in human brain structure. Nat Neurosci 20, 287–296 (2017). https://doi.org/10.1038/nn.4458

[547] Pregnancy Causes Lasting Changes in a Woman's Brain. https://www.scientificamerican.com/article/pregnancy-causes-lasting-changes-in-a-womans-brain/ [access 2020-12-06]

[548] Rice, K., Redcay, E. Spontaneous mentalizing captures variability in the cortical thickness of social brain regions, Social Cognitive and Affective Neuroscience, Volume 10, Issue 3, March 2015, Pages 327–334, https://doi.org/10.1093/scan/nsu081

[549] How Men's Bodies Change When They Become Fathers https://parenting.nytimes.com/health/fatherhood-mens-bodies [access 2020-12-06]

[550] Oxytocin. https://www.yourhormones.info/hormones/oxytocin/ [access 2020-12-06]

[551] Barba-Müller, E., Craddock, S., Carmona, S., Hoekzema, E. Brain plasticity in pregnancy and the postpartum period: links to maternal caregiving and mental health. Arch Womens Ment Health. 2019;22(2):289–299. doi:10.1007/s00737-018-0889-z

[552] Kim, P., Strathearn, L., Swain, J. E. The maternal brain and its plasticity in humans. Horm Behav. 2016;77:113–123. doi:10.1016/j.yhbeh.2015.08.001

[553] Barba-Müller, E., Craddock, S., Carmona, S., Hoekzema, E. Brain plasticity in pregnancy and the postpartum period: links to maternal caregiving and mental health. Arch Womens Ment Health. 2019;22(2):289–299. doi:10.1007/s00737-018-0889-z

[554] The Postpartum Brain. https://greatergood.berkeley.edu/article/item/postpartum_brain [access 2020-12-06]

[555] Gettler, L., Mcdade, T., Feranil, A. Kuzawa, C. (2011). Longitudinal Evidence that Fatherhood Decreases Testosterone in Human Males. Proceedings of the National Academy of Sciences of the United States of America. 108. 16194-9. 10.1073/pnas.1105403108.

[556] Mascaro, J. S., Hackett, P. D., Rilling, J. K. Differential neural responses to child and sexual stimuli in human fathers and non-fathers and their hormonal correlates. Psychoneuroendocrinology. 2014;46:153–163.

doi:10.1016/j.psyneuen.2014.04.014

[557] Edelstein, R. S., Chopik, W. J., Saxbe, D. E., Wardecker, B. M., Moors, A. C., LaBelle, O. P. Prospective and dyadic associations between expectant parents' prenatal hormone changes and postpartum parenting outcomes. Dev Psychobiol. 2017;59(1):77□90. doi:10.1002/dev.21469

[558] Kim, P., Rigo, P., Mayes, L. C., Feldman, R., Leckman, J. F., Swain, J. E. Neural plasticity in fathers of human infants. Soc Neurosci. 2014;9(5):522–535. doi:10.1080/17470919.2014.933713

[559] Bharadwaj, S., Kulkarni, G., Shen, B. (2015). Menstrual cycle, sex hormones in female inflammatory bowel disease patients with and without surgery. Journal of Digestive Diseases, 16(5), 245–255. https://doi.org/10.1111/1751-2980.12247

[560] How Men's Bodies Change When They Become Fathers https://parenting.nytimes.com/health/fatherhood-mens-bodies [accessed 2020.12.06]

[561] Bouchet, H., Plat, A., Levréro, F., Reby, D., Patural, H., Mathevon, N. Baby cry recognition is independent of motherhood but improved by experience and exposure. Proc Biol Sci. 2020 Feb 26;287(1921):20192499. doi: 10.1098/rspb.2019.2499. Epub 2020 Feb 19. PMID: 32070250; PMCID: PMC7062011.

[562] What Is Postpartum Depression? Recognizing The Signs And Getting Help https://www.npr.org/2020/01/27/800139124/what-is-postpartum-depression-recognizing-the-signs-and-getting-help?t=15812701 89375&t=1581344521026 [access 2020-12-06]

[563] Nott, P. N., Franklin, M., Armitage, C., Gelder, M. G. (1976). Hormonal changes and mood in the puerperium. Br.J.Psychiatry 128, 379–383. doi: 10.1192/bjp.128.4.379

[564] O'Hara, M. W., Stuart, S., Gorman, L. L., Wenzel, A. (2000). Efficacy of inter- personal psychotherapy for postpartum depression. Arch. Gen. Psychiatry 57, 1039–1045. doi: 10.1001/archpsyc.57.11.1039

[565] Meyer, J. H., Ginovart, N., Boovariwala, A., Sagrati, S., Hussey, D., Garcia, A., et al. (2006). Elevated mono-amine oxidase a levels in the brain: an explana- tion for the monoamine imbalance of major depression. Arch. Gen. Psychiatry 63, 1209–1216. doi: 10.1001/archpsyc.63.11.1209

[566] Barth, C., Villringer, A., Sacher, J. Sex hormones affect neurotransmitters and shape the adult female brain during hormonal transition periods. Front Neurosci. 2015;9:37. Published 2015 Feb 20. doi:10.3389/fnins.2015.00037

[567] Bromberger, J. T., Kravitz, H. M. (2011). Mood and Menopause: Findings from the Study of Women's Health Across the Nation (SWAN) over 10 Years. Obstetrics and Gynecology Clinics of North America, 38(3), 609–625. https://doi.org/10.1016/j.ogc.2011.05.011

[568] Vliegen, N., Casalin, S., Luyten, P. The Course of Postpartum Depression: A Review of Longitudinal Studies, Harvard Review of Psychiatry: January/February 2014 – Volume 22 – Issue 1 – p 1-22 doi: 10.1097/HRP.0000000000000013

[569] How Long Can Postpartum Depression Last — and Can You Shorten It? https://www.healthline.com/health/depression/how-long-does-postpartum-depression-last [access 2020-12-06]

References

[570] Postpartum depression https://www.mayoclinic.org/diseases-conditions/postpartum-depression/symptoms-causes/syc-20376617 [access 2020-12-06]

[571] Cameron, E. E., Sedov, I. D., Tomfohr-Madsen, L. M. Prevalence of paternal depression in pregnancy and the postpartum: An updated meta-analysis. J Affect Disord. 2016 Dec;206:189-203. doi: 10.1016/j.jad.2016.07.044. Epub 2016 Jul 20. PMID: 27475890.

[572] Psouni, E., Agebjörn, J.,Linder, H.. (2017). Symptoms of depression in Swedish fathers in the postnatal period and development of a screening tool. Scandinavian Journal of Psychology. 58. 10.1111/sjop.12396.

[573] Men get postnatal depression, too https://www.theguardian.com/society/2018/sep/04/fathers-men-get-postnatal-depression-too [access 2020-12-06]

[574] Symptoms of postpartum anxiety. https://www.healthline.com/health/pregnancy/postpartum-anxiety#symptoms [access 2020-12-06]

[575] Farr, S. L., Dietz, P. M., O'Hara, M. W., Burley, K., Ko, JY. Postpartum anxiety and comorbid depression in a population-based sample of women. J Womens Health (Larchmt). 2014;23(2):120-128. doi:10.1089/jwh.2013.4438

[576] Yildizi, P. D., Ayers, S., Phillips, L. The prevalence of posttraumatic stress disorder in pregnancy and after birth: A systematic review and meta-analysis. Journal of Affective Disorders Volume 208, 15 January 2017, Pages 634-645

[577] The effect of childbirth no-one talks about. https://www.bbc.com/future/article/20190424-the-hidden-trauma-of-childbirth [access 2020-12-06]

[578] Daniels, E., Arden-Close, E., Mayers, A. Be quiet and man up: a qualitative questionnaire study into fathers who witnessed their Partner's birth trauma. BMC Pregnancy Childbirth. 2020;20(1):236. Published 2020 Apr 22. doi:10.1186/s12884-020-02902-2

[579] Why we shouldn't demonize formula feeding https://www.health.harvard.edu/blog/why-we-shouldnt-demonize-formula-feeding-2018040313557 [access 2020-12-06]

[580] Wagner, E. A., Chantry, C. J., Dewey, K. G., Nommsen-Rivers, L. A. Breastfeeding concerns at 3 and 7 days postpartum and feeding status at 2 months. Pediatrics. 2013;132(4):e865-e875. doi:10.1542/peds.2013-0724

[581] Guyton, A. C., Hall, J. E. Textbook of Medical Physiology. Philadelphia, PA: Elsevier Saunders, 11th edition 2006.

[582] How Breastfeeding Actually Works: Exploring the Science of Breast Milk Feeding. https://www.medela.com.au/breastfeeding/blog/awesome-breast-milk-facts/how-breastfeeding-actually-works-exploring-the-science-of-breast-milk-feeding [access 2020-12-06]

[583] Williams Obstetrics, 24th edition. McGraw-Hill Education, New York, USA

[584] Guyton, A. C., Hall, J. E. Textbook of Medical Physiology. Philadelphia, PA: Elsevier Saunders, 11th edition 2006.

[585] Uvnäs Moberg, K., Prime, D. Oxytocin effects in mothers and infants during breastfeeding. Infant 9 (6) 2013

[586] Emory Health Sciences. "How Dads bond with toddlers: Brain scans link oxytocin to paternal nurturing: Study looks at neural mechanisms of paternal caregiving." ScienceDaily. ScienceDaily, 17 February 2017. https://www.sciencedaily.com/releases/2017/02/170217095925.htm [access 2020-12-06]

[587] How Breastfeeding Actually Works: Exploring the Science of Breast Milk Feeding https://www.medela.com.au/breastfeeding/blog/awesome-breast-milk-facts/how-breastfeeding-actually-works-exploring-the-science-of-breast-milk-feeding [access 2020-12-06]

[588] Breastfeeding. https://www.who.int/features/factfiles/breastfeeding/en/ [access 2020-12-06]

[589] Guyton, A. C., Hall, J. E. Textbook of Medical Physiology. Philadelphia, PA: Elsevier Saunders, 11th edition 2006.

[590] National Breastfeeding Month. http://www.ufmcpueblo.com/national-breastfeeding-month/ [access 2020-12-06]

[591] Guyton, A. C., Hall, J. E. Textbook of Medical Physiology. Philadelphia, PA: Elsevier Saunders, 11th edition 2006.

[592] How do people around the world celebrate periods? https://www.actionaid.org.uk/blog/news/2019/10/18/how-do-people-around-the-world-celebrate-periods [access 2020-12-07]

[593] End the stigma. Period. https://www.unwomen.org/en/digital-library/multimedia/2019/10/infographic-periods [access 2020-12-07]

[594] Sperling, M. A. (10 April 2014). Pediatric Endocrinology E-Book. Elsevier Health Sciences. pp. 485–. ISBN 978-1-4557-5973-6.

[595] Elzouki, A. Y., Harfi, H. A., Nazer, H., Stapleton, F. B., Oh, W., Whitley, R. J. (10 January 2012). Textbook of Clinical Pediatrics. Springer Science & Business Media. pp. 3681–. ISBN 978-3-642-02202-9.

[596] Elias, C. F. Leptin action in pubertal development: recent advances and unanswered questions. Trends Endocrinol Metab. 2012;23(1):9–15. doi:10.1016/j.tem.2011.09.002

[597] Frisch, R. E . Body fat, menarche, fitness and fertility. Hum Reprod 1987; 2: 521–533.

[598] Not her real name

[599] Marshall, W. A., Tanner, J. M. Variations in pattern of pubertal changes in girls. Arch Dis Child. 1969 Jun; 44(235):291-303.

[600] Sanfilippo, J., Jamieson, M. Glob. libr. women's med., (ISSN: 1756-2228) 2008; DOI 10.3843/GLOWM.10286

[601] Function. https://www.medicalnewstoday.com/articles/277177#function [access2020-12-07]

[602] Guyton, A. C., Hall, J. E. (2006). Medical physiology 11th edition. Textbook of Medical Physiology.

[603] What is a normal vaginal pH? https://www.healthline.com/health/womens-health/vaginal-ph-balance#normal-ph [2020-12-07]

[604] Sanfilippo, J., Jamieson, M. Glob. libr. women's med., 2008. https://doi.10.3843/GLOWM.10286

References

[605] Are there any other benefits? https://www.healthline.com/health/purpose-of-pubic-hair#other-benefits

[606] A bushel of facts about the uniqueness of human pubic hair https://blogs.scientificamerican.com/bering-in-mind/a-bushel-of-facts-about-the-uniqueness-of-human-pubic-hair/ [access 2020-12-07]

[607] Function https://www.medicalnewstoday.com/articles/277177#function [access 2020-12-07]

[608] Wheeler, M. D. Physical changes of puberty. Endocrinol Metab Clin N Am 20:1, 1991

[609] Adolescent pregnancy. https://www.who.int/news-room/fact-sheets/detail/adolescent-pregnancy [access 2020-12-07]

[610] Papri, F. S., Khanam, Z., Ara,S., Panna, M. B. Adolescent Pregnancy: Risk Factors, Outcome and Prevention. Chattagram Maa-O-Shishu Hospital Medical College Journal Volume 15, Issue 1, January 2016

[611] Disease burden and mortality estimates. WHO. Global health estimates 2015: deaths by cause, age, sex, by country and by region, 2000–2015. Geneva: WHO; 2016. https://www.who.int/healthinfo/global_burden_disease/estimates/en/ [access 2020-12-07]

[612] WHO. Global health estimates 2015: deaths by cause, age, sex, by country and by region, 2000–2015. Geneva: WHO; 2016.

[613] Wall-Wieler, E., Roos, L. L., Nickel, N. C. Teenage pregnancy: the impact of maternal adolescent childbearing and older sister's teenage pregnancy on a younger sister. BMC Pregnancy Childbirth 16, 120 (2016). https://doi.org/10.1186/s12884-016-0911-2

[614] Maciejewski, D. F., van Lier, P. A. C., Branje, S. J. T., Meeus, W. H. J., Koot, H. M.. A 5□Year Longitudinal Study on Mood Variability Across Adolescence Using Daily Diaries. Child development, Volume86, Issue6, November/December 2015, Pages 1908-1921

[615] Arain, M., Haque, M., Johal, L., et al. Maturation of the adolescent brain. Neuropsychiatr Dis Treat. 2013;9:449–461. doi:10.2147/NDT.S39776

[616] https://www.nimh.nih.gov/health/publications/the-teen-brain-7-things-to-know/index.shtml

[617] Arain. M., Haque, M., Johal, L., et al. Maturation of the adolescent brain. Neuropsychiatr Dis Treat. 2013;9:449–461. doi:10.2147/NDT.S39776

[618] Ibid.

[619] Arain, M., Haque, M., Johal, L., et al. Maturation of the adolescent brain. Neuropsychiatr Dis Treat. 2013;9:449–461. doi:10.2147/NDT.S39776

[620] Hagenauer, M. H., Perryman, J. I., Lee, T. M., Carskadon, M. A. Adolescent changes in the homeostatic and circadian regulation of sleep. Dev Neurosci. 2009;31(4):276–284. doi:10.1159/000216538

[621] Crowley, S. J, Cain, S. W., Burns, A. C., Acebo, C., Carskadon, M. A. Increased Sensitivity of the Circadian System to Light in Early/Mid-Puberty. J Clin Endocrinol Metab. 2015;100(11):4067–4073. doi

[622] University of Washington. "Teens get more sleep with later school start time, researchers find." ScienceDaily. ScienceDaily, 12 December 2018. <www.sciencedaily.com/releases/2018/12/181212140741.htm>.:10.1210/jc.2015-

2775

[623] Eckert-Lind, C., Busch, A. S., Petersen, J. H, et al. Worldwide Secular Trends in Age at Pubertal Onset Assessed by Breast Development Among Girls: A Systematic Review and Meta-analysis. JAMA Pediatr. Published online February 10, 2020. doi:10.1001/jamapediatrics.2019.5881

[624] Onset of puberty in girls has fallen by five years since 1920 https://www.theguardian.com/society/2012/oct/21/puberty-adolescence-childhood-onset [access 2020-12-07]

[625] Gavela-Pérez, T., Navarro, P., Soriano-Guillén, L., Garcés, C. High Prepubertal Leptin Levels Are Associated With Earlier Menarcheal Age. J Adolesc Health. 2016 Aug;59(2):177-81. doi: 10.1016/j.jadohealth.2016.03.042. Epub 2016 Jun 11. PMID: 27297138.

[626] Gamble, J. Early starters. Nature 550, S10–S11 (2017). https://doi.org/10.1038/550S10a

[627] Deardorff, J., Ekwaru, J. P., Kushi, L. K., Ellis, B. J., Greenspan, L. C., Mirabedi, A., et al. Father Absence, Body Mass Index, and Pubertal Timing in Girls: Differential Effects by Family Income and Ethnicity Journal of Adolescent Health doi:10.1016/j.jadohealth.2010.07.032

[628] Harley, K. G., Berger, K. P., Kogut, K., et al. Association of phthalates, parabens and phenols found in personal care products with pubertal timing in girls and boys. Hum Reprod 2019; 34: 109-117. doi: 10.1093/humrep/dey337.

[629] Endocrine disruptors in personal care products associated with the timing of puberty https://www.focuson-reproduction.eu/article/ESHRE-News-Annual-Meeting-2020-Harley [access 2020-12-07]

[630] Fisher, M. M., Eugster, E. A. What is in our environment that effects puberty?. Reprod Toxicol. 2014;44:7–14. doi:10.1016/j.reprotox.2013.03.012

[631] Greenspan, L. C., Lee, M. M. Endocrine disrupters and pubertal timing. Curr Opin Endocrinol Diabetes Obes. 2018;25(1):49–54. doi:10.1097/MED.0000000000000377

[632] Fertility Rate. https://ourworldindata.org/fertility-rate [access 2020-12-07]

[633] Hawkes, K., O'Connell, J. F., Jones, N. G., Alvarez, H., Charnov, E. L. Grandmothering, menopause, and the evolution of human life histories. Proc Natl Acad Sci U S A. 1998 Feb 3;95(3):1336-9.

[634] Engelhardt, S. C., Bergeron, P., Gagnon, A., Dillon, L., Pelletier, F. Using Geographic Distance as a Potential Proxy for Help in the Assessment of the Grandmother Hypothesis. Curr Biol. 2019 Feb 18;29(4):651-656.e3. doi: 10.1016/j.cub.2019.01.027. Epub 2019 Feb 7.

[635] Chapman, S. N., Pettay, J.E., Lummaa, V., Lahdenperä, M. Limits to Fitness Benefits of Prolonged Post-reproductive Lifespan in Women. Curr Biol. 2019 Feb 18;29(4):645-650.e3. doi: 10.1016/j.cub.2018.12.052. Epub 2019 Feb 7.

[636] Menopause is when you have your last menstruation and marks the end of your reproductive years

[637] Fritz, M. A., Speroff, L. Clinical Gynecologic Endocrinology and Infertility, Philadelphia PA: Lippincott Williams & Wilkins, Eighth edition 2011.

References

[638] Blanchflower, D. G. Is Happiness U-shaped Everywhere? Age and Subjective Wellbeing in 132 Countries, NBER Working Paper No. 26641, Issued in January 2020

[639] Hall, J. E. Endocrinology of the Menopause. Endocrinol Metab Clin North Am. 2015;44(3):485-496. doi:10.1016/j.ecl.2015.05.010

[640] Guyton, A. C., Hall, J. E. Textbook of Medical Physiology. Philadelphia, PA: Elsevier Saunders, 11th edition 2006.

[641] Allshouse, A., Pavlovic, J., Santoro, N. Menstrual Cycle Hormone Changes Associated with Reproductive Aging and How They May Relate to Symptoms. Obstet Gynecol Clin North Am. 2018;45(4):613–628. doi:10.1016/j.ogc.2018.07.004

[642] Kruszynska, A., Slowinska-Srzednicka, J. (2017). Anti-Müllerian hormone (AMH) as a good predictor of time of menopause. Przeglad Menopauzalny, 16(2), 47–50. https://doi.org/10.5114/pm.2017.68591

[643] Kelsey, T. W., Wright, P., Nelson, S. M., Anderson, R. A., Wallace, W. H. A validated model of serum anti-müllerian hormone from conception to menopause. PLoS One. 2011;6(7):e22024. doi:10.1371/journal.pone.0022024

[644] Steiner, A. Z., Pritchard, D., Stanczyk, F. Z., et al. Association Between Biomarkers of Ovarian Reserve and Infertility Among Older Women of Reproductive Age. JAMA. 2017;318(14):1367–1376. doi:10.1001/jama.2017.14588

[645] Allshouse, A., Pavlovic, J., Santoro, N. Menstrual Cycle Hormone Changes Associated with Reproductive Aging and How They May Relate to Symptoms. Obstet Gynecol Clin North Am. 2018;45(4):613–628. doi:10.1016/j.ogc.2018.07.004

[646] Woods, N. F., Mitchell, E.S., Smith-Dijulio, K. Cortisol levels during the menopausal transition and early postmenopause: observations from the Seattle Midlife Women's Health Study. Menopause. 2009;16(4):708□718. doi:10.1097/gme.0b013e318198d6b2

[647] Pinkerton, J. V., Guico-Pabia, C. J., Taylor, H. S. Menstrual cycle-related exacerbation of disease. Am J Obstet Gynecol. 2010;202(3):221–231. doi:10.1016/j.ajog.2009.07.061

[648] Santoro, N. Perimenopause: From Research to Practice. J Womens Health (Larchmt). 2016;25(4):332–339. doi:10.1089/jwh.2015.5556

[649] Breast changes in older women. https://www.nhs.uk/live-well/healthy-body/breast-changes-in-older-women/ [access 2020-12-07]

[650] Yazdkhasti, M., Tourzani, Z.M., Roozbeh, N. et al. The association between diabetes and age at the onset of menopause: a systematic review protocol. Syst Rev 8, 80 (2019). https://doi.org/10.1186/s13643-019-0989-5

[651] Otolorin, E. O., Adeyefa, I., Osotimehin, B. O., Fatinikun, T., Ojengbede, O., Otubu, J. O., Ladipo, O. A. Clinical, hormonal and biochemical features of menopausal women in Ibadan, Nigeria.

Afr J Med Med Sci. 1989 Dec; 18(4):251-5.

[652] Fritz, M. A., Speroff, L. Clinical Gynecologic Endocrinology and Infertility, Philadelphia PA: Lippincott Williams & Wilkins, Eighth edition 2011.

[653] Morley, J. E., Mitchell Perry, H. III. Androgens and Women at the Menopause and Beyond, The Journals of Gerontology: Series A, Volume 58, Issue 5, May 2003, Pages M409–M416, https://doi.org/10.1093/gerona/58.5.M409

[654] Ibid.

[655] Avisa, N. E., Stellatoa, R., Crawford, S., Bromberger, J., Ganz, P., Cain, V., Kagawa-Sing, M. Is there a menopausal syndrome? Menopausal status and symptoms across racial/ethnic groups. Social Science and Medicine 52 (2001) 345–356

[656] Freedman, R. R. Menopausal hot flashes: mechanisms, endocrinology, treatment. J Steroid Biochem Mol Biol. 2014;142:115–120. doi:10.1016/j.jsbmb.2013.08.010

[657] Bansal, R., Aggarwal, N. Menopausal Hot Flashes: A Concise Review. J Midlife Health. 2019;10(1):6–13. doi:10.4103/jmh.JMH_7_19

[658] Delamater, L., Santoro, N. Management of the Perimenopause. Clin Obstet Gynecol. 2018;61(3):419–432. doi:10.1097/GRF.0000000000000389

[659] Born, L., Koren, G., Lin, E., Steiner, M. A new, female-specific irritability rating scale. J Psychiatry Neurosci. 2008;33(4):344-354.

[660] Freeman, E. W., Sammel, M. D., Hui Lin, M.S., Nelson, D. B. Associations of Hormones and Menopausal Status With Depressed Mood in Women With No History of Depression. ArchGenPsychiatry.2006;63:375-382

[661] Devanand, D. P., Sano, M., Tang, M. X., Taylor, S., Gurland, B. J., Wilder, D., et al. (1996). Depressed mood and the incidence of Alzheimer's disease in the elderly living in the community. Arch. Gen. Psychiatry 53, 175–182. doi: 10.1001/archpsyc.1996.01830020093011

[662] Zandi, P. P., Carlson, M. C., Plassman, B. L., Welsh-Bohmer, K. A., Mayer, L. S., Steffens, D. C., et al. (2002). County Memory Study, Hormone replacement therapy and incidence of Alzheimer disease in older women: the Cache County Study. JAMA 288, 2123–2129. doi: 10.1001/jama.288.17.2123

[663] Allshouse, A., Pavlovic, J., Santoro, N. Menstrual Cycle Hormone Changes Associated with Reproductive Aging and How They May Relate to Symptoms. Obstet Gynecol Clin North Am. 2018;45(4):613–628. doi:10.1016/j.ogc.2018.07.004

[664] Pincott, J. Menopause Predisposes a Fifth of Women to Alzheimer's. Being female is a risk factor for Alzheimer's. Why? Scientific American, May 2020

[665] Davis, S. R., Castelo-Branco, C., Chedraui, P., Lumsden, M. A., Nappi, R. E., Shah, D., Villaseca, P. Understanding weight gain at menopause.Climacteric. 2012 Oct;15(5):419-29. doi: 10.3109/13697137.2012.707385.

[666] Women cold water swimming in Gower to help menopause. https://www.bbc.com/news/uk-wales-47159652 [access 2020-12-07]

References

[667] Cho. M. K. Use of Combined Oral Contraceptives in Perimenopausal Women. Chonnam Med J. 2018;54(3):153–158. doi:10.4068/cmj.2018.54.3.153

[668] Ibid.

[669] Kinney, A., Kline, J.,Kelly, A., Reuss, M. L., Levin, B. Smoking, alcohol and caffeine in relation to ovarian age during the reproductive years, Human Reproduction, Volume 22, Issue 4, April 2007, Pages 1175–1185, https://doi.org/10.1093/humrep/del496

[670] Kinney A, Kline J, Levin B. Alcohol, caffeine and smoking in relation to age at menopause. Maturitas. 2006 Apr 20;54(1):27-38. Epub 2005 Nov 2.

[671] Drinking Alcohol https://www.breastcancer.org/risk/factors/alcohol [access 2020-12-07]

[672] Kinney, A., Kline, J., Kelly, A., Reuss, M. L., Levin, B. Smoking, alcohol and caffeine in relation to ovarian age during the reproductive years, Human Reproduction, Volume 22, Issue 4, April 2007, Pages 1175–1185, https://doi.org/10.1093/humrep/del496

[673] Fritz, M. A., Speroff, L. Clinical Gynecologic Endocrinology and Infertility, Philadelphia PA: Lippincott Williams & Wilkins, Eighth edition 2011.

[674] Langton, C. R., Whitcomb, B. W., Purdue-Smithe, A. C., et al. Association of Parity and Breastfeeding With Risk of Early Natural Menopause. JAMA Netw Open. 2020;3(1):e1919615. doi:10.1001/jamanetworkopen.2019.19615

[675] de Vries, E., den Tonkelaar, I., van Noord, P.A. H., van der Schouw, Y. T., te Velde, E. R, Peeters, P. H. M. Oral contraceptive use in relation to age at menopause in the DOM cohort, Human Reproduction, Volume 16, Issue 8, August 2001, Pages 1657–1662, https://doi.org/10.1093/humrep/16.8.1657

[676] Wend, K., Wend, P., Krum, S. A. Tissue-Specific Effects of Loss of Estrogen during Menopause and Aging. Front Endocrinol (Lausanne). 2012;3:19. doi:10.3389/fendo.2012.00019

[677] Yao, P., Bennett, D., Mafham, M., et al. Vitamin D and Calcium for the Prevention of Fracture: A Systematic Review and Meta-analysis. JAMA Netw Open. 2019;2(12):e1917789. doi:10.1001/jamanetworkopen.2019.17789

[678] Wend, K., Wend, P., Krum, S. A. Tissue-Specific Effects of Loss of Estrogen during Menopause and Aging. Front Endocrinol (Lausanne). 2012;3:19. doi:10.3389/fendo.2012.00019

[679] Menopause Predisposes a Fifth of Women to Alzheimer's. https://www.scientificamerican.com/article/menopause-predisposes-a-fifth-of-women-to-alzheimers/ [access 202107]20

[680] https://www.scientificamerican.com/article/menopause-predisposes-a-fifth-of-women-to-alzheimers/

[681] The Colorado thyroid disease prevalence study. Canaris GJ, Manowitz NR, Mayor G, Ridgway EC Arch Intern Med. 2000 Feb 28; 160(4):526-34.

[682] Wend, K., Wend, P., Krum, S. A. Tissue-Specific Effects of Loss of Estrogen during Menopause and Aging. Front Endocrinol (Lausanne). 2012;3:19. doi:10.3389/fendo.2012.00019

[683] Raz, R. Urinary tract infection in postmenopausal women. Korean J Urol. 2011;52(12):801–808. doi:10.4111/

kju.2011.52.12.801

[684] The menopause myth: how demonised HRT came back from the brink. https://www.theguardian.com/society/2020/feb/09/the-menopause-myth-how-demonised-hrt-came-back-from-the-brink [access 2020-12-07]685 https://www.theguardian.com/society/2020/feb/09/the-menopause-myth-how-demonised-hrt-came-back-from-the-brink

[686] Jiang, Y., Tian, W. The effects of progesterones on blood lipids in hormone replacement therapy. Lipids Health Dis. 2017;16(1):219. doi:10.1186/s12944-017-0612-5

[687] Barth, C., Villringer, A., Sacher, J. (2015). Sex hormones affect neurotransmitters and shape the adult female brain during hormonal transition periods. Frontiers in Neuroscience, 9(FEB), 1–20. https://doi.org/10.3389/fnins.2015.00037

Vocabulary

Androgens,
"male" hormones like testosterone. They develop the primary male sex organs and secondary sex characteristics at puberty. Synthesized in the testes, the ovaries, and the adrenal glands. They are precursors to oestrogens and function in libido and sexual arousal.

Anterior pituitary gland,
part of the pituitary gland (hypophysis). Regulates stress, growth, breastfeeding and reproduction. The latter by adjusting the LH and FSH.

Anti-Müllerian Hormone (AMH)
is needed for oestrogen production. Is available in the ovaries and can be used as a marker of ovarian reserve and PCOS.

Body mass index (BMI),
body weight divided by the square of the body height (m^2).

Bisphenol-A (BPA),
a chemical that hardens plastic and makes it transparent, a suspected endocrine disruptor.

Cholesterol,
a precursor for the biosynthesis of cortisol, aldosterone, progesterone, oestrogens, and testosterone, and vitamin D.

Clomiphene,

a drug that induces ovulation.

Corpus luteum,

what is left of the follicle after ovulation, produces progesterone

Corticotropin-releasing hormone (CRH)

is released by the hypothalamus or the placenta, regulates the blood flow between the placenta and the foetus. Promotes bonding. Lead to increased cortisol.

Cortisol,

stress hormone. Late in pregnancy it is about three times as high as in non-pregnant women.

Eicosanoids,

are "local hormones" such as prostaglandins.

Endocrine system,

is the collection of glands that produces hormones in the body. Glands in the brain: hypothalamus, pituitary gland, and pineal gland plus adrenal glands, the pancreas, the testes and the ovaries.

Endometriosis,

a condition when endometrial cells grow outside uterus.

Dopamine,

"feel-good chemical".

Fallopian tube,

connects the ovary to the uterus, the egg and sperm meets here.

Feedback loops,

means that hormones are released by glands in the brain and stimulate growth of cells in other parts of the body. Thereafter another inhibiting hormone is produced.

Female sexual hormones,
  most typical are oestrogen and progesterone.

Fertile phase,
  is the three to seven days leading up to and including ovulation.

Follicular phase,
  is the phase before ovulation, and the period after the luteal phase.

Follicle stimulating hormone (FSH)
  turns the follicles into eggs.

Gonadotropin-releasing-hormone (GnRH)
  produced by hypothalamus, releases gonadotropins, (among others the follicle-stimulating-hormone (FSH) and luteinizing hormone (LH)).

Human chorionic gonadotropin (hCG)
  prevents degradation of corpus luteum during the early pregnancy. Detected in home pregnancy test. Also injected during IVF.

Human placental lactogen,
  a pregnancy hormone that among others changes the mother's metabolism

Hypothalamic-pituitary-ovary axis (HPO axis),
  coordinates hormonal changes in the menstrual cycle.

Hypothalamic-pituitary-adrenal (HPA) axis,
  regulates how we react to stress.

Hypothalamic-pituitary-thyroid (HPT) axis,
  regulates metabolism.

Hyperprolactinemia,
  the pituitary gland produces too much prolactin.

Hyperthyroidism,
  the thyroid gland produces too much hormone.

Hypothyroidism,

the thyroid produces too little hormone.

IVF,

in vitro fertilization.

Insulin resistance,

cells fail to respond normally to insulin.

Laparoscopy,

insertion of a thin tube with a high-intensity light and a camera.

Leptin,

regulates energy balance by inhibiting hunger,

Luteinizing hormone (LH),

triggers ovulation of an egg. LH starts to rise around 34-36 hours before ovulation. Used to identify the fertile phase.

Luteal insufficiency,

means insufficiently built uterine wall.

Luteal phase,

begins with formation of the corpus luteum and ends in either pregnancy or next menstruation.

MRI scan,

magnetic resonance imaging.

Müllerian ducts/tract,

in foetus develop into the fallopian tubes, the uterus, and the vagina.

Neurotransmitters,

promote communication between the nerve cells and the target cells.

Oestrogen,

produced in the ovaries and in fat cells. Generates sexual desire, protects heart, brains, and bones, creates more blood vessels. Even men have oestrogen.

Oxytocin,

"bonding" hormone, induces labour.

Phytoestrogen,

natural in plants. Mimics the body's own oestrogen. The most common is soy.

Polycystic ovary syndrome (PCOS),

a hormonal disorder that can lead to irregular cycles and excess of androgens.

Premenstrual syndrome (PMS),

physical and/or emotional changes at a specific period of the menstrual period.

Premenstrual dysphoric disorder (PMDD),

stronger mental symptoms for a longer time than for PMS.

Progesterone,

promotes pregnancy, build the wall of the uterus so the egg can implant and start growing, relaxes smooth musculature and acts calming.

Progestin,

synthetic medication with effect as progesterone, used in hormonal birth control (the pill and IUD, intrauterine device).

Prolactin,

stimulates breastfeeding.

Prostaglandins,

heals injured cells by provoking fever, pain, and inflammation. In women, they act in menstruation (contract uterus), and initiate ovulation (weakening the wall of the follicle) and labour pains.

Relaxin,

relaxes the ligaments and the bones in the pelvis.

Serotonin,

makes us feel happy.

Testosterone,

produces in the testes, male sex hormone but women have it too.

Thyroid Stimulating Hormone (TSH)

Wolffian ducts,

here the connections between the testicles and other male genital parts are created

CPSIA information can be obtained
at www.ICGtesting.com
Printed in the USA
LVHW080757251121
704359LV00021B/373